Milton Aparecido Deperon Junior

Tillage and compaction implements for maize cultivation

Milton Aparecido Deperon Junior

Tillage and compaction implements for maize cultivation

ScienciaScripts

Imprint

Any brand names and product names mentioned in this book are subject to trademark, brand or patent protection and are trademarks or registered trademarks of their respective holders. The use of brand names, product names, common names, trade names, product descriptions etc. even without a particular marking in this work is in no way to be construed to mean that such names may be regarded as unrestricted in respect of trademark and brand protection legislation and could thus be used by anyone.

Cover image: www.ingimage.com

This book is a translation from the original published under ISBN 978-613-9-66113-8.

Publisher:
Sciencia Scripts
is a trademark of
Dodo Books Indian Ocean Ltd. and OmniScriptum S.R.L publishing group

120 High Road, East Finchley, London, N2 9ED, United Kingdom
Str. Armeneasca 28/1, office 1, Chisinau MD-2012, Republic of Moldova, Europe
Printed at: see last page
ISBN: 978-620-7-98450-3

DEDICATORY

To my family (Milton, Mary, Mariana, Grandpa Ângelo and Mrs Meiry), for their teachings, love, affection, etc. To my beloved wife Andréia Deperon for her love, affection and encouragement, above all for her patience and understanding from the beginning of this work.

ACKNOWLEDGEMENTS

My supervisor Prof Nelci Olszevski for her immense support, trust, dedication and friendship in the development of this work, helping and guiding the writing as well as the results. Thanks to you, this achievement was possible. Thank you very much.

To my colleague Hideo de Jesus Nagahama for his friendship, commitment and unquestionable support in the development of this work, and thanks for his help in writing up and analysing the data. Thanks to you, this achievement was possible. Thank you very much.

To Prof Jorge Wilson Cortez for his support, trust, ideas and teachings in carrying out this work. Thank you very much.

To the Professors of the Postgraduate Programme in Agricultural Engineering at UNIVASF, for their knowledge and patience during the classes. Thank you very much.

To my fellow postgraduate students for the laughs and fun times we had together, especially Fernando and Alberto. Thank you very much.

To the staff of CCA-UNIVASF, for their immense support in the development of agricultural activities, as well as to Prof Helder Ribeiro Freitas for providing the experimental area. Thank you very much.

To everyone I may be forgetting, who directly or indirectly helped me to carry out this work. Thank you very much.

SUMMARY

As a result of the intensification of agriculture, whether under conventional sowing methods or with the use of no-till farming, there has been an increase in compacted areas. The aim of this study was to assess the influence of soil preparation implements and compaction levels on the physical attributes of a typical dystrophic yellow Argissolo with a sandy texture and its possible impact on the agronomic characteristics of maize. The treatments consisted of three soil preparation implements and four levels of compaction caused by tractor traffic. A randomised block experimental design was used in a sub-divided plot layout with four replications, with the soil preparation implements in the plots and the compaction levels in the sub-plots. The following soil physical attributes were assessed in the 0.00 - 0.10; 0.10 - 0.20; 0.20 - 0.30 and 0.30 - 0.40 m layers: soil density (Ds), total porosity (Pt) and soil mechanical resistance to penetration (RP). The following assessments were carried out on the plants: plant height (AP), first ear insertion height (APE), stem diameter (DC), number of grains (NG), first ear rows (FE), first ear length (CE), first ear diameter (DE), plant dry matter (MSP), root dry matter (MSR), 1000 grain mass (MG) and yield (P). Soil preparation implements influenced DS and Pt in the 0.00-0.10 m layer, but RP was not affected. The compaction levels led to an increase in DS in the 0.00 - 0.10 m and 0.30 - 0.40 m layers, Pt in the 0.00 - 0.10 m layer and PR in the 0.00 - 0.10 m; 0.10 - 0.20 m and 0.20 - 0.30 m layers. Of the crop's agronomic characteristics evaluated, soil preparation implements only influenced AP, APE and DC, with higher values for AP and APE being observed when the harrow was used for soil preparation. The increase in PR in the 0.00 - 0.30 m layer caused by compaction levels restricted all the crop characteristics evaluated, with the exception of CE, DE and MG. It was possible to observe that PR values between 1.53 and 3.33 MPa caused reductions in AP, APE, DC, NG, FE, MSP and P of 19, 14, 15, 20, 11, 39 and 22 per cent respectively. MSR increased quadratically up to a PR of 2.18 MPa, and from this to 3.33 MPa there was a 53 per cent reduction. This showed the importance of checking soil compaction conditions to prevent losses in the maize crop.

Keywords: Soil density; Mechanical strength; *Zea mays.*

CHAPTER 1

INTRODUCTION

The replacement of native forests by cultivated areas causes alterations in the chemical, physical and biological characteristics of soils as a result of the change from a natural condition to an anthropised one. These changes occur mainly in modern, technologised agriculture, due to the use of machinery and implements that promote intense soil turning processes and, in many cases, these operations are responsible for negative changes in soil quality.

There are currently three main tillage systems: conventional tillage, reduced tillage and no-till tillage. Conventional tillage can be characterised by more aggressive tillage where harrowing and ploughing are generally used. Reduced tillage aims to reduce tillage and the main implements used in this system are scarifiers and subsoilers, which are considered less aggressive as they work the soil in a different way to harrows and ploughs. No-till farming, on the other hand, is characterised by minimal tillage, which only takes place in the sowing rows. Thus, tillage operations are carried out with the aim of improving and creating favourable conditions for crop germination and root growth. However, humidity conditions during tillage, the soil's clay and organic matter content, tillage depth and the type of implement used can lead to changes in the soil's structure, restricting root growth (DE MARIA et aL, 1999).

As a result of the intensification of agriculture, be it conventional sowing or the use of no-till farming, there has been an increase in compacted areas and in these, sometimes intense variations along the soil profile. However, soil compaction, from an agronomic point of view, only has significance when it interferes with the process of plant growth and development (COLLARES, 2005).

Various physical and mechanical soil parameters are used to measure compaction, as well as plant analyses such as root and aerial development and productivity. There is intense research into the limits or ranges of soil parameters, such as soil density, that are critical to plant development, and the search for parameters that are representative of soil compaction and make it possible to compare the most different soils and managements, such as the "degree of compaction" or "relative compaction" (SUZUKI, 2005).

Due to the great economic importance of the maize crop, which stands out among cereals as one of the most widely planted crops in the world, and in Brazil has the second largest cultivated area with approximately 15.8 million hectares planted in the 2012/2013 harvest (CONAB, 2014), several authors have studied the effects of tillage systems and compaction on root dry matter production (KLEPKER, 1991; FOLONI et al, 2003; FREDDI, 2007), plant dry matter production

(KLEPKER, 1991; FREDDI, 2007), productivity (TREIN, 1988; KLEPKER, 1991; CENTURION and DEMATTÊ, 1992; GAGGERO, 1998; MELLO IVO and MIELNICZUK, 1999; SILVA et aL, 2000a; SEIXAS, 2001; SUZUKI, 2005; FREDDI, 2007; FREDDI et aL, 2009; PENEDO, 2011), plant height (TREIN, 1988; SEIXAS, 2001; SUZUKI, 2005; FREDDI, 2007), among other agronomic aspects. However, most of these authors carried out their work mainly on Latosols, which are found more frequently in the Cerrado and southern regions of the country. The São Francisco Valley region has great potential for growing maize because it has an extensive irrigated perimeter, the possibility of crop rotation, a dry climate that guarantees a low incidence of leaf diseases, high luminosity, among other aspects.

The aim of this work was to assess the influence of tillage implements and compaction levels on the physical attributes of a typical dystrophic yellow Argissolo with a sandy texture and their possible impact on the agronomic characteristics of maize.

CHAPTER 2

LITERATURE REVIEW

2.1 Soil compaction

Agricultural soils have suffered major disturbances, with compaction being identified as the main cause of these changes due to the traffic of tractors and agricultural machinery under inadequate management conditions (RICHART, et aL, 2005).

Camargo and Alleoni (1997) define soil compaction as an alteration in the arrangement of its constituent particles. For Lima (2004), the term soil compaction refers to the process that describes the decrease in volume of unsaturated soils when a certain external pressure is applied, which can be caused by the traffic of agricultural machinery, transport equipment or animals.

Richart et al (2005) report that the compaction process depends on external and internal factors. The external factors are characterised by the type, intensity and frequency of load applied, while the internal factors are stress history, moisture, texture, structure, initial soil density and carbon content.

According to Richart et aL (2005), agricultural machinery traffic is the main cause of soil compaction, which has been intensified by the modernisation of agriculture, with an increase in the mass of equipment and the intensity of land use. The same author reports that this process has not been accompanied by a proportional increase in tyre size and width, resulting in significant changes to the soil's physical properties. According to Silva et aL (2000b), the tyres usually used on tractors and harvesters sold in Brazil have rigid sidewalls and are called diagonal tread tyres. This rigidity prevents the tyre from moulding to the ground according to the irregularities of the terrain and, as a result, its contact area is reduced, increasing the pressure on the soil surface.

Radford et aL (2000) observed a significant increase in soil density at a depth of 0.11 m and in the cone index in the 0.00 - 0.11 and 0.13 - 0.18 m layers when a machine with a mass of 10 and 2 Mg on the front and rear axles, respectively, travelled once on a Vertisol. Collares (2005), working on a Red Argissolo with a sandy loam texture, observed that the additional compaction caused by the traffic of a machine with a mass of 10 Mg increased soil density values and reduced total porosity. Chan et al. (2006) found in a sodic Vertisol higher density and penetration resistance and lower macroporosity, in addition to reduced root growth of canola and wheat, as well as reduced canola productivity in the lines of machine traffic.

According to Lanças (2002), the use of machines such as ploughs, harrows and rotary hoes solves the problem of soil compaction in the surface layers, but in most cases transfers it to deeper layers. The use of these machines, almost always at the same soil preparation depth and for several consecutive years, has contributed to the appearance of compacted layers just below the line of action of the active parts of the machines, known as subsurface compaction. Centurion and Demattê (1985), studying conventional tillage (ploughing, heavy harrowing + light harrowing); reduced tillage (heavy harrowing + light harrowing); indirect sowing (mowing and herbicide application) and super tillage (two ploughs + heavy harrowing + light harrowing), concluded that the reduced, conventional and super tillage systems induced the formation of compacted layers at depths of 0.10, 0.20 and 0.20 m respectively. According to Richart et al. (2005), subsoil compaction can be alleviated by subsoiling and tends to be permanent due to the peculiarities of subsoiling.

Soil texture plays a major role in the compaction process (RICHART et al., 2005). According to Reichert et al. (2003), compaction occurs more intensely in clayey soils; however, these soils are more resistant to disintegration, while sandy soils have fewer compaction problems but are highly susceptible to disintegration. Moura et al. (2008), working with artificial compaction in two soils with different textural classes, found that the highest degree of compaction, represented by a soil density of 1.98 Mg m^{-3} in a Planossolo with 0.055 kg kg^{-1} of clay, did not restrict the development of radish bulbs, while a different behaviour was observed in a Nitossolo with 0.39 kg kg^{-1} of clay, where at a soil density of 1.47 Mg m^{-3} there was no development of radish bulbs. These results corroborate Lima (2004), who emphasises that at the same soil density and water potential, the soil is more compressible the higher the clay content and the lower the organic substance content. According to Kaiser (2010), clay soils, because they are more porous, can facilitate root growth, even under conditions of high resistance. Sandy soils, because they are less porous and offer greater friction between particles, with less variation in humidity, can restrict root growth. According to Dias Júnior and Miranda (2000), this can be explained by the fact that the particles in soils with a predominant sand fraction rearrange themselves more acutely than in soils with a predominant clay fraction.

Another important factor that influences the degree of compaction is soil moisture which, according to Dias Júnior and Pierce (1996), for the same condition is the property that governs the amount of deformation that can occur in the soil at the time of mechanised operations. Thus, when soils are drier, their load-bearing capacity may be sufficient to withstand the pressures applied and soil compaction may not be significant. However, under conditions of high humidity, the soil deforms more easily and compacted layers form (SWAN et aL, 1987 *apud* RICHART et aL, 2005). Secco (2003), in work carried out on a dystrophic red latosol with a clay texture and a typical dystrophic red latosol with a clay texture, found that there was an average increase in preconsolidation pressure as density increased and the soil suffered a reduction in the degree of water saturation from 91% to 58% and from 93% to 68% respectively, suggesting that the drier and more compacted the soil, the greater its bearing capacity, with the particles and/or aggregates being more cohesive. According to the author, for a degree of saturation of less than 45% and 60%, respectively, for the first and second soils, the pre-consolidation pressure was little influenced by the soil's humidity and density. In a moist soil, according to the author, water acts as a lubricant between the particles, making the soil softer, altering its state of consistency and consequently reducing the soil's load-bearing capacity. According to Smith et aL (1997) *apud* Imhoff (2002), this process continues until water saturates practically all the soil pores. From then on, a new increase in the soil's water content will not respond with an increase in density, since the water cannot be compressed. Silva et aL (2006) found a reduction in root dry matter production and root density of *Eucalyptus urophylla* due to the compaction of an oxide-gibbsite Red-Yellow Latosol in a pot, at a moisture content of 0.20 kg kg^{-1} , compared to the other moistures (0.05 and 0.10 kg kg^{-1}). In a kaolinitic Yellow Latosol, there was only a reduction in root density with compaction at humidities of 0.10 and 0.20 kg kg^{-1} .

The conditions needed to avoid compaction are often difficult to match because some of them, such as humidity, depend on climatic conditions. However, it is necessary to plan and organise activities to take these factors into account, thus avoiding compaction, a complex problem that is difficult to recover from (REICHERT et aL, 2007). Oliveira et al. (2003), found that for a typical dystrophic Red Latosol with a clay texture, at lower tensions, i.e. higher humidity, the uniaxial compression tests were higher, showing the need to monitor soil humidity when making decisions about the entry of machinery into agricultural areas. Similar results were found by other authors (BRAIDA, 2004; DIAS JÚNIOR et aL, 2004; LIMA et al., 2006). According to Albuquerque et al. (2001), moisture is one of the factors that influences the level of soil compaction, so it is recommended to avoid using heavy machinery and to remove animals from the area when the soil moisture is above the friability point in a crop-livestock integration system on a Nitossolo to avoid compaction.

According to Reichert et al. (2007), among the properties used to assess soil compaction, density is perhaps the safest, as it has less or no dependence on other factors, such as moisture.

Changes in soil density are easily measured and can be an indicator of changes in soil quality and ecosystem functioning. Increases in density generally imply a decrease in soil quality for root growth, reduced aeration and undesirable changes in the behaviour of water in the soil, such as reduced infiltration (SILVA, 2008).

Soil resistance to root penetration and soil density are related to the state of soil compaction and many studies have sought values that cause restrictions on plant root development and reduced productivity. The difficulty lies in isolating the effect of these physical properties and soil moisture; therefore, doubts persist as to which soil property best characterises the state of compaction and which is sensitive to variations in soil management (SILVA, 2003).

According to Silva (2008), water content and soil density influence penetration resistance. Penetration resistance increases with soil compaction and as the soil dries out. Therefore, the effect of density on root growth is more intense in relatively dry soils. In wet soils, greater density is needed to restrict root penetration. For example, a traffic-compacted layer with a density of 1.6 Mg m^{-3} can completely restrict root penetration in very dry soil and allow root penetration in moist soil.

According to De Maria et al. (1999), to characterise compaction between soil layers, penetration resistance gives better results than soil density. However, as resistance is dependent on soil moisture and density, it is difficult to determine critical or restrictive values or ranges, and it is easier to obtain these values for properties such as soil density (REICHERT et al, 2007).

The assessment of soil compaction, whether through density or penetration resistance, shows a good relationship with root growth. Generally, with an increase in density or penetration resistance, there is a reduction in root development (REICHERT et al., 2007). According to Reichert et al. (2003), maximum compaction is generally around 0.05 metres deep, with soil density values of around 1.5 Mg m^{-3} and macroporosity varying between 5% and 8%. Some soil density values that could cause severe impediments have been indicated in the literature, although they are strictly empirical. Critical soil density values have been found by several authors (SILVA and KAY, 1997; KLEIN, 1998; TORMENA et al., 1998, TORMENA et al., 1999; IMHOFF et al., 2001; SILVA, 2003; SILVA et al, 2004) when the optimum water interval was equal to zero, using these data Reichert et al. (2003) proposed critical values according to textural classes, which were: 1.25 to 1.3 Mg m^{-3} for very clayey soils; 1.3 to 1.4 Mg m^{-3} for clayey soils; 1.4 to 1.5 Mg m^{-3} for loamy soils; 1.56 Mg m^{-3} for silty loam soils and 1.7 to 1.8 Mg m^{-3} for sandy loam soils.

Cintra and Mielniczuk (1983) found that soil density of 1.30 Mg m^{-3} and soil resistance to penetration of 1,100 kPa reduced the root system of several crops by 50 per cent in a very clayey purple latosol. Alvarenga et al. (1996) determined that in a very clayey dark red latosol, the critical density of the soil for the root development of various legumes was around 1.25 Mg m^{-3}, for crotalaria juncea *(Crotalaría juncea)*, pigeonpea *(Canavalia ensiformes)* and Ceará pigeonpea *(Canavaeia brasiliensis)*, and above 1.35 Mg m^{-3}, for guandu *(Cajanus cajari)*. De Maria et al. (1999) observed in a very clayey purple latosol that a soil density of 1.21 Mg m^{-3} reduced the root growth of soya beans. Secco (2003), working on a dystrophic Red Latosol with a clayey texture, observed that a soil density of 1.62 Mg m^{-3} reduced wheat yields, and on a typical dystrophic Red Latosol with a clayey texture, a soil density of 1.54 Mg m^{-3} reduced wheat and maize yields. Collares (2005) observed that a soil density of 1.74 Mg m^{-3} in a dystrophic Argissolo Vermelho with a sandy loam texture reduced the root development of the bean plant. Silva et al. (2000a), in work carried out on a Red-Yellow Podzolic, observed that the density of maize roots correlated inversely with soil density in the 0.10 - 0.25 m layer. Freddi et al. (2009), studying the effects of different compaction treatments on a typical dystrophic red latosol, found that macroporosity and soil resistance to penetration were the variables most sensitive to changes caused by tractor traffic, while soil density was little altered, showing a lower coefficient of variation.

Soil resistance to penetration has been identified as one of the limiting factors for crop development and establishment, as this characteristic expresses the degree of compaction, varying with the type of soil and the species grown, and its causes have been attributed to agricultural machinery traffic (RICHART et aL, 2005). For some authors (CANARACHE, 1990; PABIN et aL, 1998 *apud* REICHERT et al., 2007) penetration resistance is dependent on soil moisture, density and particle size distribution. Therefore, a dry or denser soil has greater resistance compared to a moist or less dense soil, while for the same humidity, a clay soil has greater resistance than a sandy soil.

Tormena et al. (1998) and Guimarães et al. (2002) concluded that in compacted soils, plant development is lower and this was attributed to the mechanical impediment to root growth, which resulted in a smaller volume of soil explored, less absorption of water and nutrients and consequently lower productivity of the crops evaluated. According to Abreu et al. (2004), the limiting factor for crop productivity is not always the soil's mechanical resistance, but rather a set of factors, such as: the soil's own resistance to root penetration, the air space used for gas exchange and the amount of water available to the plants.

In the literature, different penetration resistance values have been presented as critical or restrictive to plant development and productivity (REICHERT et al., 2007). Taylor et al. (1966), working under laboratory conditions, indicated that a penetration resistance of 2.0 MPa was restrictive. In another study, also under laboratory conditions, Canarache (1990) suggested some penetration resistance limits considering root development: values lower than 2.5 MPa would not limit root growth; values between 2.6 and 10 MPa would cause some limitations and values higher than 10 MPa would not allow root growth. Sene et al. (1985) consider critical penetration resistance values ranging from 6.0 to 7.0 MPa for sandy soils and around 2.5 MPa for clayey soils. This variation may be associated with the type of soil, the species or variety involved and the soil humidity at the time of assessment.

Meroto Jr. and Mundstock (1999), working in pots, found that a resistance of 2.0 MPa caused a minimal reduction in the dry matter of roots, aerial part and root length of wheat, while for a resistance of 3.5 MPa the restrictions were severe. Beutler and Centurion (2004a) found values of 2.30 and 2.90 MPa to be restrictive for the production of dry matter in the aerial part of rice, respectively, in a typical medium-textured dystrophic red latosol and typical clay-textured eutrophic red latosol. These same authors also found a reduction in rice crop yields when the soil resistance to penetration was 2.38 and 2.07 MPa, when these soils had water contents of 0.14 kg kg^{-1} and 0.27

kg kg⁻¹ respectively. Beutler et al. (2004), working with soya and rice crops, found that penetration resistance of 1.66 and 2.22 MPa for soya and 0.27 and 2.38 MPa for rice, respectively, for water contents of 0.11 kg kg⁻¹ and 0.14 kg kg⁻¹ , reduced the productivity of these crops in a typical medium-textured dystrophic red latosol in a pot. According to the authors, with an increase in water content, a higher value of penetration resistance can be tolerated, and the lower critical value of penetration resistance with a reduction in water content may be indicative of an interaction with the effect of water potential in the soil, restricting the plant's physiological activity.

In field conditions, De Maria et al. (1999) found that a penetration resistance of 2.09 MPa could be determining a reduction in the root growth of soya beans in a very clayey dystrophic Latossolo Roxo. Secco et al. (2009) found that penetration resistance values of 2.65 and 3.26 MPa, respectively in a dystrophic Red Latosol with a clayey texture and in a dystrophic Red Latosol with a clayey texture, were not restrictive for soya crops under field conditions; however, for maize crops, the penetration resistance value of 3.26 MPa was high enough to reduce the crop's grain yield. According to the authors, the maize crop was more sensitive to the compaction conditions found in the two latosols, showing that grasses, in comparison to legumes, were more susceptible to the negative effects on the soil's physical attributes imposed by compaction conditions. Beutler et al. (2006), in a study carried out on a medium-textured red latosol with four soya cultivars, observed that productivity decreased from soil penetration resistance values of 2.24 to 2.97 MPa depending on the cultivars studied.

2.2 Soil preparation systems

Soil preparation can be defined as mechanical manipulation with the aim of loosening and mixing the soil, eradicating weeds, incorporating crop residues and fertilisers, and creating an optimum degree of compaction for root growth (GIL and VANDEN BERG, 1968). A long time ago, the planting of any crop required soil preparation, which almost always involved the use of ploughs and harrows.

pulverised the soil, the better the preparation was considered to be (INOUE, 2003). Nowadays, this thinking has changed, given the development of research relating soil attributes to agricultural quality and productivity.

From a technical point of view, the tillage system should contribute to maintaining or improving the quality of the soil and the environment, as well as achieving satisfactory crop yields in the long term (COSTA et aL, 2003).

According to Dorneles (2011), tillage methods can vary from those that cause intense soil mobilisation, such as conventional tillage, to so-called conservation tillage, in which the soil is little mobilised or worked only in the sowing line, as is the case with no-till farming.

Conventional soil preparation consists of mechanically disturbing the surface layers, generally using one ploughing and two harrowing operations (FASINMIRIN and REICHERT, 2011), to reduce compaction, increase pore spaces and thus increase permeability and air and water storage, as well as facilitating the development of plant roots. According to Carvalho Filho et al. (2007), ploughing and harrowing operations are called primary and secondary preparation respectively. Periodic primary soil preparation is characterised by deeper operations, carried out mainly by disc or mouldboard ploughs, with the aim of cutting, raising and inverting the soil bed. During this preparation, vegetation and crop remains are incorporated. Secondary tillage involves operations carried out by harrows, mainly with smooth, jagged or smooth and jagged discs, with the aim of levelling and loosening the soil, incorporating herbicides, correctives and fertilisers, and eliminating weeds (ALVARENGA et al., 2002; BENTIVENHA et al., 2003).

The no-till system is characterised by soil disturbance only in the sowing furrow, with crop rotation and the maintenance of straw on the soil surface (BEUTLER et al., 2007). As a result, it reduces erosion processes due to the absence of tillage and the presence of vegetation cover, improves the soil's structural and biological conditions, increases its water infiltration and retention capacity and its organic matter content, reduces soil temperature variations and reduces water loss through evaporation, thus promoting environmental preservation and increasing agricultural productivity (FASINMIRIN and REICHERT, 2011). However, this system tends to lead to the formation of compacted layers in the soil caused by the traffic of machinery and implements, causing changes to the soil structure (CAMARA and KLEIN, 2005). As a result of the practices used and the length of time they are used, this can result in an increase in soil density and, consequently, soil compaction, one of the main physical problems that affects water dynamics and limits crop productivity (CAMARA and KLEIN, 2005; BEUTLER et al., 2007).

As tillage systems work the soil in different ways, various studies have been carried out to check the changes caused to the physical and chemical properties and to the crops. Falleiro et al. (2003), in a study carried out on a Cambrian Yellow Red Argisol, found that no-till farming produced higher soil density values when compared to the other treatments. Corsini and Ferraudo (1999), in work carried out on epieutrophic purple latosol with a clayey texture, concluded that subsoiling with ploughing and harrowing, operations present in conventional soil preparation, increased the porosity of the topsoil as well as the potential for root development, while no-till farming, in the first three agricultural years, decreased the porosity and root development of the topsoil. Boukounga (2009), studying the physical properties of a typical dystrophic red Argissolo with a sandy loam texture, subjected to different tillage systems (no-till, conventional tillage and reduced tillage), observed that soil density in the 0.00 - 0.10 m layer was significantly lower when disc ploughs and harrow graders were used than when planting was used.

Trein (1988) found better results for maize yields on a red Argissolo when it was ploughed and harrowed. Dorneles (2011) found statistically superior results for soya yields on a typical dystrophic red Argissolo with a sandy loam texture, when it was subjected to conventional soil preparation, while for maize, the best results were found in the no-till system, which did not differ statistically from the other systems. Fontanela (2012), studying the effects of conventional and no-till systems on the productivity of sugarcane and cassava crops, observed in a typical Argissolo Vermelho-Amarelo dystrophic with a sandy loam texture that, in the sugarcane crop, no-till provided the highest productivity (112,68 Mg ha^{-1}) with no statistical difference from conventional tillage (94.43 Mg ha^{-1}). For cassava, no-till farming also provided the highest yield (32.7 Mg ha^{-1}) with no statistical difference from conventional tillage (24.3 Mg ha^{-1}).

The different implements available for preparing the soil cause changes in its chemical, physical and biological properties. Each implement works the soil in its own way, altering these properties in different ways (SÁ, 1998). The main implements used in soil preparation are ploughs, harrows, subsoilers and scarifiers.

Ploughing is an operation to invert layers of soil, bringing benefits such as: soil aeration; better penetration, movement and water retention; breaking down and incorporating organic matter and green manures; weed control and incorporation of fertilisers and correctives. Ploughs can be classified in various ways, including according to their active organ, which can be either disc or mouldboard (GALETI, 1988). Mouldboard ploughs promote better inversion of the soil bed and have greater penetration capacity, inverting the soil layers with less spalling effect. Crop residues are deposited at the bottom of the ploughed layer and are little mixed with the soil. Disc ploughs work better in more adverse conditions, but the ley is inverted on smaller slopes and the soil crumbling effect is greater. Crop residues remain closer to the surface and are well mixed with the soil (BALASTREIRE, 1990 *apud* FONTANELA, 2012). Stone and Moreira (1999) studied the effects of different tillage systems on soil compaction and bean compartment in a study carried out on a dark red, clay-loamy latosol under irrigation. The authors observed that when the mouldboard plough was used, lower penetration resistance values and a more uniform distribution of the root system along the soil profile were found; however, this did not lead to higher bean yields when compared to the other tillage systems.

According to Stolf et al. (2008), the most common method of tillage is carried out initially with medium (intermediate) and/or heavy (plough) harrows to decompress the first layer of soil and cut the plant material. This operation is complemented by the levelling harrow with the aim of loosening and levelling the soil surface, finalising the operation. In addition, according to Galeti (1988), harrowing can also be used to eliminate weeds, cut crop residues, bury seeds, fertilisers and correctives and carry out erosion control for the construction of mechanical soil conservation practices. Viana et al. (2006) report that heavy harrows, like other disc implements, cause greater compaction at shallower depths because the total weight of the implement is distributed over a very small area of the disc. Cortez et al. (2011), in a study carried out on a typical dystrophic yellow Argissolo with a sandy texture using three types of harrow, found no significant differences in soil density in the layers studied depending on the implements used, while total porosity values were significantly lower when a harrow with 0.61 m diameter discs and a scarifier were used, which did not differ from each other. These authors also concluded that the mechanical resistance to penetration showed adequate values up to the working depth of the implements and was high in the layers below, except for the preparation carried out with the scarifier.

Scarification is a soil preparation technique that offers minimal tillage, keeping cultural remains on the soil surface and acting as one of the alternatives often recommended to reduce the effects of soil compaction and, consequently, reduce density and increase porosity (FASINMIRIN and REICHERT, 2011). Scarifiers contain rods that are used in soil management and have advantages over disc implements in that they do not promote an inversion of layers, thus obtaining greater operational capacity and, above all, less alteration of the soil structure. They are used to prepare the soil and break up compacted sub-surface layers, thus facilitating root penetration and water infiltration into the soil (FONTANELA, 2012). Camara and Klein (2005) found that scarification reduced soil density and led to greater water infiltration, saturated soil hydraulic conductivity and surface roughness than no-till on a typical dystrophic red latosol with a clay texture. They found no significant difference between the managements for the parameters total porosity and soil macroporosity. In a study carried out on a dystrophic purple latosol with a very clayey texture, which was subjected to five primary tillage treatments: heavy harrow/heavy harrow;
scarifier/scarifier; scarifier/levelling harrow; scarifier/direct seeding and direct seeding/direct seeding De Maria et al. (1999) found that the use of the scarifier resulted in greater profile uniformity and lower density and resistance values, also indicating greater soil porosity, even in combination with the direct seeding system in winter.

2.3 Maize cultivation

Maize (Zea *mays) is* a crop belonging to the *Poaceae* family. Its monoecious character and characteristic morphology result from the suppression, condensation and multiplication of various parts of the basic anatomy of grasses. The vegetative and reproductive aspects of the maize plant can be modified through interaction with environmental factors that affect the control of developmental ontogeny (MAGALHÃES et aL, 1994).

In the 2012/2013 harvest, Brazil produced 81 million tonnes of maize on an area of 15.8 million hectares, and according to estimates by CONAB (2014) a slight reduction in area of around 2% and 2.5% in production is expected for the 2013/2014 harvest. According to CONAB's (2014) estimate for the 2013/2014 harvest, the North and Northeast regions will see a 20.1% increase in production from 6.5 million tonnes to 7.8 million tonnes.

According to Salviano et aL (1980) *apud* Choudhury et aL (1991), the exploitation of maize in irrigated areas contributes to the implementation of a crop rotation programme, aimed at reducing the incidence of pests and diseases, using the residual effect of fertilisers and the use of crop remains for animal feed, due to their high nutritional value.

Carvalho et aL (2012) evaluated the performance of maize hybrids in the northeastern region of Brazil in two networks of trials, the first of which encompassed 15 environments and the second 13 environments under rainfed cultivation. For the first network of trials, they found average yield results varying between 7.52 Mg ha^{-1} and 7.40 Mg ha^{-1} for the second network, thus demonstrating good production potential for the north-eastern region of Brazil.

According to Aldrich et aL (1975) *apud* Boller et aL (1998), an ideal sowing bed for planting maize should provide good conditions for germination and initial root development of the seedlings, provide adequate weed control, allow the smooth operation of sowing and cultivation machinery, preserve or improve soil aggregation and allow the maximum possible infiltration of water.

In the case of maize, the results for the different soil management systems are also very different. Higher maize yields in the no-till system compared to other soil management systems were reported by Hernani (1997) and Ismail et aL (1994) *apud* Kluthcouski et aL (2000), and lower yields by Oliveira et aL (1989) and Balbino et aL (1994). Mello Ivo and Mielniczuk (1999), in a study carried out on Dark Red Argissolo with three tillage systems (Conventional, Reduced and Direct), found no significant differences in the productivity of the maize crop as a function of the systems.

Preparation with a mouldboard plough is rarely used because it requires more time and energy to operate than the others, although it can result in higher yields of maize, soya and wheat (BALBINO; OLIVEIRA, 1992; KOCHHANN; DENARDIM, 1997; KLUTHCOUSKI et aL, 2000, *apud* YOKOYAMA et aL, 2002), when compared to the no-till system or preparation with a plough harrow. This is due to the inferior development of the root system in these types of tillage, due to soil compaction in the surface or subsurface layer, respectively (YOKAYAMA et aL, 2002). Kluthcouski et aL (2000), in a study involving four types of soil management (no-till (PD); deep scarification (EP); plough harrow (GA) and deep ploughing with mouldboards (AP)) and three levels of fertiliser in an eutrophic purple latosol with a loamy sandy loam texture, found significant yields for ploughing compared to the other types of management, followed by deep scarification, plough harrow and no-till. Yokayama et aL (2002) also observed better grain yields in maize when a perforated red latosol was prepared with a mouldboard plough, which on average over the six years of evaluation was 9% and 7% higher than those obtained with harrow preparation and no-till, respectively. Corsini and Ferraudo (1999), in work carried out on an epieutrophic purple loam with a clayey texture subjected to different cultivation systems, observed that subsoiling with ploughing and harrowing increased the porosity of the topsoil, as well as the root development potential of the maize crop.

Rosolem et al. (1994) observed that increasing the density of the soil from 1.03 to 1.72 Mg.m^{-3} increased the resistance to penetration from 0.05 to 2.0 MPa, causing total impediment to the growth of maize roots. Freddi (2007), studying the effect of additional compaction on a medium-textured red latosol, observed that just one pass with a 4.0 Mg tractor at the water content equivalent to field capacity was enough to raise the soil density above 1.46 Mg m^{-3} , limiting the grain yield of the maize crop.

Mechanical penetration resistance can restrict the root development of maize, and various studies have been carried out to determine critical limits for crop development. Tavares Filho et al. (2001) stated that a penetration resistance of 3.5 MPa in a loamy purple latosol did not restrict the root development of maize, but it did influence its morphology. Foloni et al. (2003) in a study of two maize materials in a medium-textured red distroferric latosol, found that for both materials soil compaction had a negative influence on root and aerial growth, although soil penetration resistance of around 1.4 MPa prevented the maize root system from developing in depth. Freddi (2007) in a medium-textured Red Latosol found that increasing the soil's mechanical resistance to penetration above 1.65 MPa restricted the agronomic characteristics and productivity of maize and for a clay-textured Eutroferric Red Latosol when the soil's mechanical resistance to penetration reached 1.16 MPa, maize productivity was significantly lower. Freddi et al. (2009) in a study carried out on a typical medium-textured dystrophic red latosol, reported that the yield of two maize hybrids was significantly lower when the soil penetration resistance reached 2.15 MPa.

CHAPTER 3

MATERIAL AND METHODS

3.1 Location

The experiment was conducted in Petrolina-PE, at the Campus de Ciências Agrárias (CCA) of the Universidade Federal do Vale do São Francisco - UNIVASF, located at the geographical coordinates of 9°19'10" south latitude and 40°33'39" west longitude, with an average altitude of 376 metres. The area used for the experiment had been fallow since 2006. According to Brasil (1973), using the Köppen classification, the climate of the region where the experimental area was located is tropical semi-arid, type BSwh' characterised as very hot, semi-arid steppe type with scarce and irregular rainfall. The climatic data for the region during crop development is shown in Figure 1.

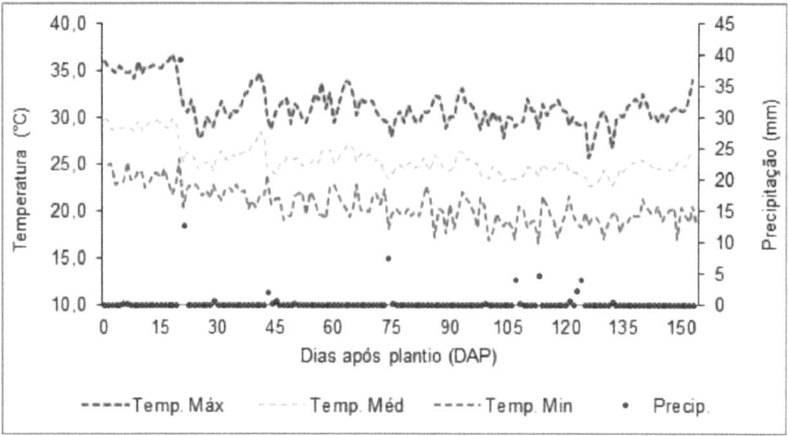

Figure 1. Climatic data for the region during crop development. Source: UNIVASF/CCA.

The soil where the experiment was carried out is classified as Argissolo Amarelo Distrófico típico textura arenosa according to Amaral et al. (2006), using the Brazilian Soil Classification System (EMBRAPA, 2006). The composition

The granulometry and gravimetric soil moisture (US) at the time of soil preparation, application of compaction levels and measurement of soil mechanical resistance for the 0.00 - 0.10; 0.10 - 0.20; 0.20 - 0.30 and 0.30 - 0.40 m layers are shown in Table 1.

Table 1. Granulometric composition, gravimetric soil moisture (US) at the time of soil preparation, application of compaction levels and measurement of soil mechanical resistance at the depths of the Yellow Argissolo.

Depth	Clay	Sand	Silt	US: Soil preparation	US: Compactness levels	US: Mechanical resistance
m	--------- kg kg^{-1} -------------			------------------------- kg kg $^{-1}$ ------------------------		
0,00-0,10	0,09	0,88	0,03	0,112	0,088	0,078
0,10-0,20	0,10	0,88	0,02	0,096	0,088	0,090
0,20 - 0,30	0,08	0,85	0,07	0,088	0,080	0,075
0,30 - 0,40	0,14	0,81	0,05	0,071	0,082	0,079

The chemical analysis of the soil in the area where the experiment was carried out is shown in Table 2.

Table 2. Chemical composition of Argissolo Amarelo in the 0.00 - 0.20 m layer.

MO	PH	C.E.	P	K	Ca	Mg	In	Al	H + Al	S	CTC	V
g kg^{-1}	H O_2	dS m^{-1}	mg dm^{-3}	-------------------- emole dm^{-3}								%
6,52	5,9	0,48	48	0,33	1,5	0,6	0,01	0,05	1,32	2,44	3,76	65

The area where the experiment was set up has a fixed-mesh conventional sprinkler irrigation system, which was used to maintain soil humidity at the time of preparation, application of compaction, data collection and crop water maintenance. Crop irrigation followed the volumes suggested by Andrade et al. (2006) according to the phenological phase and evaporative demand, so approximately 590 mm were applied throughout the crop cycle.

3.2 Inputs and equipment

On 1st March 2013, the soil preparation implements were used and on 27th March 2013, the levels were applied.

For this purpose, the tractor's second gear was used, corresponding to a speed of 3.88 km h^{-1} . Before the compaction levels were applied, the entire experimental area was harrowed. The equipment described in Table 3 was used to carry out the agricultural activities.

Table 3. Agricultural equipment used in the experiment.

Equipment	Brand	Model	Features
Levelling harrow	TATU Marchesan	GH	Lightweight *tandem* harrow with 7 discs of *0.51 m* diameter (20") in each of the four sections, 14 of which are cut at the front and 14 smooth at the rear. Working width of *2.62 metres* and working depth of *0.06-0.12 metres*. Discs spaced *0.19 m* apart. Implement mass of 519 kg.
Ploughing harrow	TATU Marchesan	ACTOR	Lightweight *off-set* harrow with 7 cut-out discs of *0.61 m* diameter (24") in each of the two sections. Working width of *1.5 m, and* working depth between *0.10 - 0.18 m.* Discs spaced *0.23 m* apart. Implement mass 1465 kg.

			Reversible mouldboard plough, with two *0.41 m* long (16") cut-out mouldboards. Working width of approximately *1.0 m, and* working depth of between *0.2 - 0.4 m.* Implement mass of 570 kg.
Mouldboard plough	Maschietto	ARH^2	
Scarifier	TATU Marchesan	AST	Scarifier with 3 shanks spaced *0.34 m* apart, narrow *0.05 m* tip. Working width of *0.7 m and* working depth of up to *0.35 m.* Implement mass of 292kg.
Seeder - Fertiliser	TATU Marchesan	T^2 SI	4 rows with a maximum spacing of *0.9 metres and* a minimum of *0.45 metres.* Implement mass of 675 kg.
Tractor	Valtra	785 TDA	Power of 55.2 kW (75 hp) with 12.4 - 24 R1 front tyres and 18.4 - 30 R1 rear tyres. Equipment mass of approximately *3.5 tonnes.*

The maize sown in the area was the single hybrid DKB390PRO, whose main characteristics are: 845 degree days to flowering; average plant height of 2.2 - 2.4 m; ear insertion height of 1.25 - 1.4 m; semi-hard yellow-orange grains; intended for grain production. Sowing was carried out mechanically on 28 March 2013, using a fertiliser seeder set to distribute 7.6 seeds per metre. When the crop reached the V3 stage (MAGALHÃES and DURÃES, 2006), thinning was carried out to achieve a final population of 6 plants per metre.

Fertilisation was based on the recommendation made by CFSEMG (1999) for grain yields of 6 - 8 Mg ha^{-1} and was carried out as follows: sowing fertiliser consisted of 400 kg ha^{-1} of formula 6-24-12 (NPK) containing 6% Ca and 6% S, with the following amounts of nutrients applied: 24 kg ha^{-1} of N; 96 kg ha^{-1} of P2O5; 48 kg ha^{-1} of K2O; 24 kg ha^{-1} of Ca and 20 kg ha^{-1} of S. Top dressing was applied using 150 kg ha^{-1} of the formula 20-0-20 (NPK) at 15, 25 and 35 days after emergence (DAE). In total, the following amounts of each nutrient were applied: 114 kg ha^{-1} of N; 96 kg ha^{-1} of P O$_{25}$; 138 kg ha^{-1} of K$_2$ O; 24 kg ha^{-1} of Ca; 20 kg ha^{-1} of S.

To determine Soil Moisture (US), Soil Density (Ds) and Total Porosity (Pt), metal rings with a capacity of 128 cm were used^3 , as well as a drying oven with digital temperature control and an electronic digital balance with a maximum capacity of 5000 g and a minimum of 200 g and precision of 0.01g.

To determine the soil's mechanical resistance to penetration (RP), an impact penetrometer model IAA/Planalsucar developed by Stolf et al. (1983) was used, with the following characteristics: plunger mass of 4 kg, free fall stroke of 0.4 m, cone with a diameter of 0.0128 m and a solid angle of 30° and rod with a diameter of approximately 0.01 m.

To determine Plant Height (PH) and Height of Insertion of the First Ear (HEI), a measuring stick with a height of 3.0 metres and an accuracy of 0.05 metres was used. A digital caliper with a precision of 0.01 mm was used to determine thatch diameter (CD), ear length (EL) and ear diameter (ED).

To determine root dry matter (RSM), an adapted cylindrical monolith with a volume of 2.1 dm^3, 2.0 mm sieves, a drying oven with digital temperature control and an electronic digital balance with a maximum capacity of 5000 g and a minimum of 200 g and precision of 0.01 g were used. To determine plant dry matter (DM), a drying oven with digital temperature control and an electronic digital balance with a maximum capacity of 5000 g and a minimum of 200 g and precision of 0.01 g were used. To determine the mass of 1000 grains (MG) and yield (PG), an electronic digital scale with a maximum capacity of 5000 g and a minimum of 200 g and precision of 0.01 g was used.

3.3Treatments   and experimental design

The treatments consisted of three types of tillage implement (IP_1 = plough harrow; IP_2 = mouldboard plough and IP_3 = scarifier) and four levels of compaction (NC_0 = soil not tilled; NC_3 = 3 passes of a 3.5 Mg tractor; NC_6 = 6 passes of a 3.5 Mg tractor and NC_9 = 9 passes of a 3.5 Mg tractor). The SP_2 and SP_3 treatments were previously harrowed in order to incorporate the spontaneous plants. To simulate the levels of compaction, the tractor drove over the entire experimental plot so that the tyres compressed areas parallel to each other. The number of times the tractor travelled varied according to the treatment, with the traffic overlapping the previous one so that the entire area of each plot was travelled an equal number of times.

The experimental design used was a randomised block design (RBL) with four replications, where the soil preparation implements were placed in the plots and the compaction levels in the sub-plots.

Each experimental plot covered an area of 12.6 m^2 and was made up of four rows of maize spaced 0.9 m apart and 3.5 m long, with the two centre rows considered useful. Therefore, the useful area of the plot was 6.3 m^2 .

3.   4Evaluations

3.4.   1Physical attributes of   the soil

On 7th April 2013, in the crop inter-row, at one point per sub-plot, the following soil physical attributes were assessed in the 0.00 - 0.10; 0.10 - 0.20; 0.20 - 0.30 and 0.30 - 0.40 m layers: US, Ds, Pt and RP.

The US was determined using the gravimetric method (EMBRAPA, 2011). Deformed samples were randomly collected from the experimental area. Equation 1 shows how US was calculated.

$$US \; = \; \frac{Mu - MS}{Ma - MS} * 100 \qquad\qquad \text{(equação 1)}$$

in which,

US: Soil moisture (%),

Mu: Wet mass of the soil (g),

Ms: Soil dry mass (g), Ma: Ring mass (g).

The volumetric ring methodology was used to determine Ds, following the recommendations of EMBRAPA (2011). Equation 2 shows how DS was calculated.

$$Ds = \frac{Ms}{V}$$ (equação 2)

(equation 2)
in which,

Ds: Soil density (Mg m^{-3}),

Ms: Soil dry mass (Mg),

V Volume of the ring (m^3).

Pt was determined using the methodology described by Camargo et al. (1986). Equation 3 shows how Pt was calculated.

$$Pt = \frac{(Vt - Vs)}{Vt} * 100$$ (equação 3)

in which,

Vs = Vt - Vv and Vv = (Msat - Ms)

Pt: Total porosity (m^3 m^{-3}),

V t: Total volume (cm^3),

V s: Volume of solids (cm^3),

V v: Volume of voids (cm^3),

Msat: Mass of saturated soil (g), Ms: Mass of dry soil (g).

The RP was determined according to the instructions presented by Stolf et al. (1983). The data obtained was transformed into MPa according to the methodology adapted from Stolf (1991).

$$RP = 0{,}56 + 0{,}689N$$ (equação 4)

in which,

RP: Mechanical resistance of the soil to penetration (MPa)

N: Number of impacts

3.4.2 Agronomic characteristics of the crop

The following evaluations were carried out: Plant height (AP), First ear insertion height (APE), Stalk diameter (DC), Number of grains (NG), First ear rows (FE), Ear length (CE), Ear diameter (DE), Plant dry matter (MSP), Root dry matter (MSR), Mass of 1000 grains (MG) and Productivity (P). The AP, APE and DC evaluations were carried out at the R1 stage (Styles-stigmas are visible), according to the scale proposed by Magalhães and Durães (2006).

Six plants per plot were used in the useful area to determine the AP, APE and DC. The AP was determined from ground level to the insertion of the flag leaf. The SPL was determined from ground level to the insertion point of the first spike. CD was determined using the average of two measurements of the second internode after ground level (FREDDI, 2007).

The experiment was harvested on 15 July 2013. Six plants per sub-plot in the useful area were used to determine NG, FE, CE, DE, MG and MSR. NG was determined by counting the grains from the ears sampled. To determine FE, the number of rows of ears sampled was counted. EC was determined by measuring the distance between the base of the ear and its apex. To determine ED, the measurement was taken in the centre of the ear. NG was used to determine MG. These were weighed and the values obtained transformed (simple rule of three) into the weight of 1000 grains. To determine the MSR, roots were collected at a depth of 0.15 metres using the monolith method (BOHN, 1979). The roots were separated from the soil by washing with a water jet and then taken to an oven with forced ventilation at 65°C until they reached a constant weight. PG and MSP were determined on 1 m^2 of the useful area of each sub-plot and the values found were extrapolated to one hectare. P was obtained by weighing the grain mass and considering a standard humidity of 13%. To determine MSP, whole plants were used and taken to a forced ventilation oven at 65°C until they reached constant weight.

3.5 Analysing the data

The data on the soil's physical attributes was subjected to analysis of variance according to the layers studied, and when statistical significance was observed at 5% by the F test, the Tukey test at 5% was used to compare means. The agronomic characteristics of the crop were also subjected to analysis of variance, and when statistical significance was observed at 5% by the F test, comparisons of means were made for the soil preparation implements using the Tukey test at 5% and regressions were adjusted for the levels of compaction. Person correlation analyses were carried out between DS and PR and their values were submitted to the t-test at 5%. Statistical analyses were carried out using the SISVAR programme (FERREIRA, 2000).

CHAPTER 4

RESULTS AND DISCUSSION

4.1 Physical attributes of the soil

4.1.1 Soil density (Ds)

Ds values were influenced by tillage implements and compaction levels ($p<0.05$), and the interaction between these factors was not significant, so their effects were studied separately according to the soil layers (Table 4).

Table 4 F-test for soil density (Mg m^{-3}) as a function of tillage implement and compaction level.

F-test	Soil layers (m)			
	0,00-0,10	0,10-0,20	0,20 - 0,30	0,30 - 0,40
IP	27,12*	0,58^{ns}	2,71^{ns}	0,50^{ns}
NC	7,50*	0,96^{ns}	0,99^{ns}	3,10*
IP x NC interaction	1,75^{ns}	0,65^{ns}	1,05^{ns}	1,23^{ns}
C.V. (%) - IP	1,60	5,88	2,95	5,93
C.V. (%) - NC	4,89	3,62	3,11	3,88

ns = not significant; * = significant at 5% probability; C.V. = Coefficient of variation (%); IP = Soil preparation implements; NC = Compaction levels.

The Ds values for the soil preparation implements and for the compaction levels, considering all the layers studied, ranged from 1.64 to 1.83 Mg m^{-3} , higher than those found by Cortez et al. (2011), which ranged from 1.30 to 1.43 Mg m^{-3} in a study carried out on the same type of soil. Although the results found are close to the critical Ds values proposed by Reichert et al. (2003), of 1.70 to 1.80 Mg m^{-3} for sandy loam soils, they may not cause damage to crops, as observed by Streck (2003), who found that in an arenic dystrophic red Argissolo, only Ds greater than 1.8 Mg m^{-3} and macroporosity less than 0.10 m^3 m^{-3}

reduced plant height, leaf area and bean yield by around 50 per cent.

In the 0.00 - 0.10 m layer, statistically different Ds values were observed for the different tillage implements, with the highest value being observed in SP2 (1.76 Mg m^{-3}), which did not differ from SP3 (1.73 Mg m^{-3}) (Figure 4). SP1 provided the lowest Ds value (1.69 Mg m^{-3}), which can be attributed mainly to the fact that ploughing harrows destroy aggregates, leaving the soil looser. In addition, this lower Ds occurred in the 0.00 - 0.10 m layer, because the implement worked the soil

"

at a depth that varied between 0.10 - 0.18 m. In the other layers, it was not possible to observe any differences in DS depending on the tillage implements used. Feitosa et al. (2013), in a study carried out on sandy-textured Yellow Argissolo, found no statistical differences in DS between the tillage and soil layers assessed; however, the lowest DS values observed in the 0.00 - 0.10 m layer were found when the soil was subjected to harrowing similar to that used in this study. Mazurama et al. (2011), in a study carried out on Argissolo Vermelho Amarelo with a sandy loam texture, found lower Ds values for the 0.03 - 0.12 m layer when the soil was subjected to preparation involving harrows.

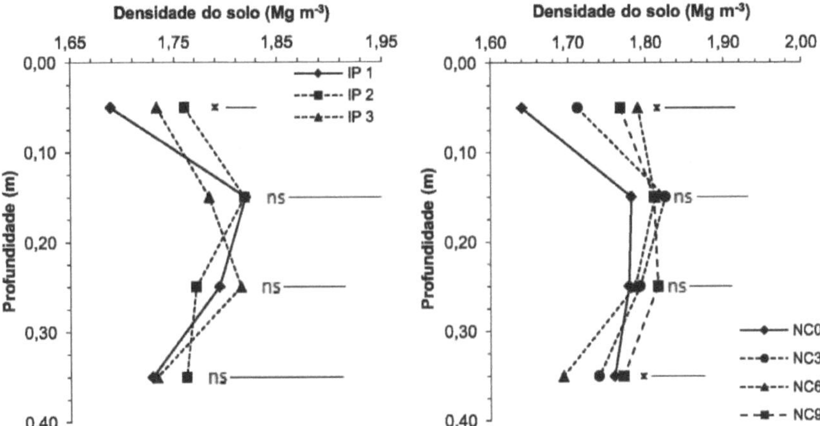

Figure 2. Soil density as a function of tillage implements (IP1 = plough harrow; IP_2 = mouldboard plough and IP_3 = scarifier) and compaction levels (NC_0 = soil not tilled; NC_3 = 3 passes of a 3.5 Mg tractor; NC_6 = 6 passes of a 3.5 Mg tractor and NCg = 9 passes of a 3.5 Mg tractor). The lines in each layer represent the MSD value; ns = not significant; * = significant at 5% by Tukey's test.

The DS values were statistically different for the compaction levels only in the 0.00 - 0.10 m and 0.30 - 0.40 m layers, and no differences could be seen in the other layers (Figure 4). In the 0.00 - 0.10 m layer, NCo provided an average Ds value (1.64 Mg m⁻³) that was lower than those obtained in NCe (1.79 Mg m⁻³) and NCg (1.78 Mg m⁻³) and statistically equal to NC3 (1.71 Mg m⁻³). NC3, NCe and NC9 generated Ds values considered critical, according to Reichert et al. (2003), which could interfere with the crop's production. In the 0.30 - 0.40 m layer, the averages observed in NCo, NC3 and NCe (1.73, 1.74 and 1.69 Mg m⁻³ respectively) were statistically equal by the Tukey test, with NCe differing from NC9 (1.79 Mg m⁻³), the latter being statistically equal to the first two (Figure 4). Streck et al. (2004), in a study carried out on Argissolo Vermelho Amarelo with a sandy loam texture, subjected to additional compaction by machine traffic, also found an increase in Ds as a function of the number of passes and observed for the 0.05 - 0.10 m layer that after four passes, the Ds value was statistically higher in relation to the non-trafficked soil. Similar results were also found in the 0.08 - 0.11 m layer by Beutler et al. (2009) in arenic Yellow Red Argisol.

4.1.4 Total porosity (Pt)

The Pt results showed statistically different values ($p<0.05$) only in the 0.00 - 0.10 m layer as a function of tillage implements and compaction levels, and their interaction was not significant (Table 5).

Table 5 F-test for Total porosity ($m^3\ m^{-3}$) as a function of tillage implements and compaction levels.

F-test	Soil layers (m)			
	0,00-0,10	0,10-0,20	0,20 - 0,30	0,30 - 0,40
IP	12,42*	$0,70^{ns}$	$0,24^{ns}$	$0,54^{ns}$
NC	7,36*	$1,05^{ns}$	$0,23^{ns}$	$0,47^{ns}$
IP x NC interaction	$1,68^{ns}$	$0,68^{ns}$	$0,70^{ns}$	$1,45^{ns}$
C.V. (%) - IP	4,88	12,01	5,93	10,12
C.V. (%) - NC	10,2	10,19	7,81	8,89

ns = not significant; * significant at 5% probability; C.V. = Coefficient of variation (%); IP = Soil preparation implements; NC = Compaction levels.

In the 0.00 - 0.10 m layer, the highest Pt was found in IP3, which had a statistically higher average than the others (Figure 3). Mazurana et al. (2011), working on Argilosso Vermelho Amarelo with a sandy loam texture with different tillage implements, found higher Pt values in the 0.12 - 0.20 m layer when scarifiers were used. For these authors, this was due to an increase in the volume of macropores, since micropores were destroyed by the action of the scarifiers. According to Schaefer et al. (2001), the effect of tillage on soil porosity values may not be very evident, with effects on the shape and distribution of pores along the profile being more common. For the same type of soil, Cortez et al. (2011) and Feitosa et al. (2013) found much higher Pt values than those presented in this study, possibly due to the initial soil conditions and the management used, which led to lower Ds, a property that has a close and inverse relationship with Pt, resulting in higher values.

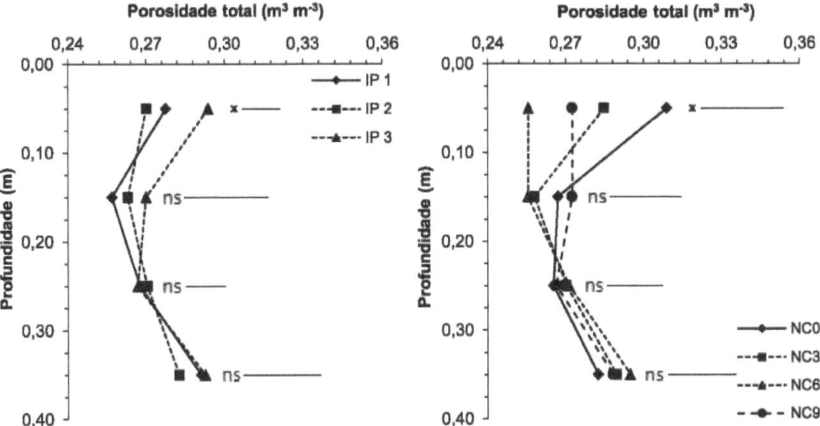

Figure 3 Total porosity as a function of tillage implement (IP$_T$ = plough harrow; IP$_2$ = mouldboard plough and IP$_3$ = scarifier) and compaction levels (NC$_0$ = soil not tilled; NC$_3$ = 3 passes of a 3.5 Mg tractor; NC$_6$ = 6 passes of a 3.5 Mg tractor and NCg = 9 passes of a 3.5 Mg tractor). The lines in each layer represent the MSD value; ns = not significant; * = significant at 5% by Tukey's test.

As with the tillage implements, Pt was only influenced by compaction levels in the 0.00 - 0.10 m layer, with no statistical differences (p<0.05) for the other layers (Figure 3). For the 0.00 - 0.10 m layer, the NCo (non-trafficked soil) showed the highest Pt value, which was statistically higher than the NCe and NCg treatments, the latter two being statistically equal.

The tractor used to simulate compaction, because it had a mass of 3.5 Mg, did not reduce Pt in the deeper layers, as observed by Streck et al. (2004) who, in a study carried out on Argissolo Vermelho Amarelo with a sandy loam texture, used a 10 Mg machine to simulate additional compaction and found significant differences in Pt up to 0.30 m depth. According to Salire et al. (1994), this is explained by the fact that compaction below the arable layer is a function of the total axle load and compaction on the surface is a function of tyre inflation pressure.

4.1.5 Mechanical resistance of the soil to penetration (RP)

PR is one of the main physical attributes used to check the degree of soil compaction, which directly and indirectly affects crop development. Only compaction levels provided significant increases (p<0.05) in PR values (Table 6).

Table 6. F-test for soil mechanical resistance to penetration (MPa) as a function of tillage implements and compaction levels.

F-test	Soil layers (m)			
	0,00-0,10	0,10-0,20	0,20 - 0,30	0,30 - 0,40
IP	2,93^{ns}	4,24^{ns}	1,90^{ns}	1,06^{ns}

NC	12,95*	10,44*	6,81*	2,66^{ns}
IP x NC interaction	0,43^{ns}	0,67^{ns}	0,60^{ns}	1,12^{ns}
C.V. (%) - IP	25,06	26,60	46,38	29,66
C.V. (%) - NC	43,58	33,56	32,20	27,36

ns = not significant; * significant at 5% probability; C.V. = Coefficient of variation (%); IP = Soil preparation implements; NC = Compaction levels.

For Taylor et al. (1966) and Silva et al. (1994), a value of 2.0 MPa becomes restrictive for the growth of roots and aerial parts of plants, while Sene et al. (1985) consider PR values greater than 6.0 MPa for sandy soils and 2.5 MPa for clay soils to be restrictive. Values higher than 2.0 MPa were found due to the soil preparation implements used in this work (Figure 4).

Although there were no statistical differences, contrasting behaviours were observed in the 0.00 -0.10 and 0.10 - 0.20 m layers, depending on the tillage implements. In the 0.00 - 0.10 m layer, IP3 had a higher PR value than the other implements, while the opposite occurred in the 0.10 - 0.20 m layer. Possibly the action of the scarifier shanks caused the compacted layer to break up, reducing the PR. Similar results were found by Mazurana et al. (2011) on Argissolo Vermelho Amarelo with a sandy loam texture at depths of 0.10 m and 0.20 m, where it was found that the action of the scarifier fitted with a breaker roller reduced PR by 100 and 90 per cent, respectively, when compared to the direct sowing system. Collares (2005) also found that scarification was effective in reducing compaction up to a depth of 0.20 m in Argissolo Vermelho Amarelo with a sandy loam texture.

It was possible to observe a good correlation between DS and PR in the 0.00 - 0.10, 0.10 - 0.20 and 0.20 - 0.30 m layers as a function of compaction levels (Figure 4). According to Kaiser (2010), soil resistance to penetration has a direct relationship with density and an inverse relationship with soil moisture and these are the factors that control its intensity. In his work on Argissolo Vermelho Amarelo with a sandy loam texture, he found a better correlation between US and RP. Rosa (2009) in his work on Argissolo Vermelho Amarelo with a loamy sandy loam texture and Kamimura (2008) in his work on Argissolo Vermelho with a sandy loam texture observed a linear tendency for PR to increase as a result of the increase in Ds.

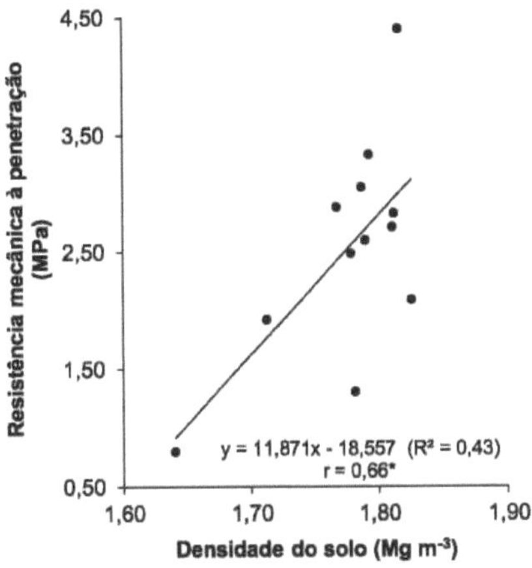

Figure 4 - Relationship between soil density (Ds) and soil mechanical resistance to penetration (RP) of the Yellow Argissolo.

Machine traffic caused an increase in PR up to a depth of 0.30 m and no statistical differences could be observed between the compaction levels below this (Figure 5). For this reason, the average PR values of the 0.00 - 0.30 m layer were used in the discussion with the attributes related to the maize crop. Similar results were found by Beutler et al. (2009) in arenic Yellow Red Argisol subjected to different traffic intensities, where increases in PR were observed up to the 0.25 m layer.

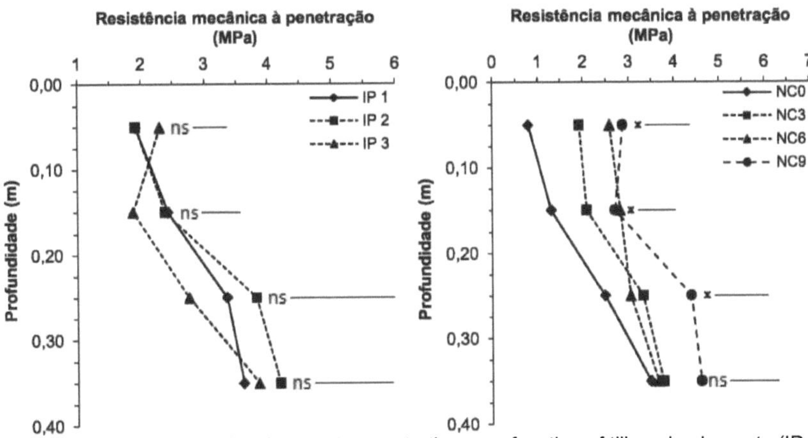

Figure 5. Soil mechanical resistance to penetration as a function of tillage implements (IP₁ = plough harrow; IP₂ = mouldboard plough and IP₃ = scarifier) and compaction levels (NC₀ = non-trafficked soil; NC₃ = 3 passes of a 3.5 Mg tractor; NC₆ = 6 passes of a 3.5 Mg tractor and NC₉ = 9 passes of a 3.5 Mg tractor), ns = not significant; * = significant at 5% by Tukey's test.

For the 0.00 - 0.10 m layer, NCo provided the lowest PR value (0.79 MPa) and was statistically different from the others, while NC3 (1.92 MPa) had a statistically similar average to NCe (2.60 MPa) and NC9 (2.88 MPa). Only NCo and NC3 had PR values lower than 2.0 MPa, and tractor traffic led to increases in PR of 143, 229 and 264% respectively for NC3, NCe and NC9. Figure 5 shows that the greatest variations in PR values were found in the 0.00 - 0.10 m layer, in agreement with the results found by Streck et al. (2004) in Argissolo Vermelho Amarelo with a sandy loam texture and by Bergamin et al. (2010) in Latossolo Vermelho with a very clayey texture. According to Botta et al. (2002) and Rosa (2009), high PR values in the subsurface layer can be attributed to compaction generated by traffic and induced by repeating the number of passes in the same place.

In the 0.10 - 0.20 m layer, only in the non-trafficked soil (NCo = 1.30 MPa) were PR values lower than 2.0 MPa, which did not differ statistically from the treatment subjected to 3 passes of the 3.5 Mg tractor (NC3 = 2.09 MPa), and differed from the other treatments (NCe = 2.82 MPa, NC9 = 2.70 MPa). For this soil layer, tractor traffic led to an increase in RP of 61,
117 and 108% respectively for 3, 6 and 9 passes. Rosa (2009) in a study on Argissolo Vermelho Amarelo with a sandy loam texture, found that traffic (4 and 8 passes of a machine with a mass of 10 Mg) generated higher PR values in the 0.05 - 0.15 m layer, regardless of the tillage system used.

For the other layers studied, all the compaction levels imposed gave PR values higher than 2.0 MPa, even when the soil was not trafficked. This leads to two possible hypotheses: the first is pedogenetic, as it is a characteristic of Yellow Argisols to thicken in depth due to the higher clay content (textural gradient), and the second is that the implements used to prepare the area did not work the soil below the 0.20 m layer. As this area has been fallow for several years, this could be its initial condition. For the 0.20 - 0.30 m layer, NCo, NCe and NC3 had statistically equal averages (2.48, 3.05 and 3.33 MPa respectively), with NC9 having the highest PR value (4.40 MPa). For this layer, the increase in PR due to tractor traffic was 41, 34 and 77 per cent respectively for 3, 6 and 9 passes. For the 0.30 - 0.40 m layer, there were no statistical differences depending on the compaction levels imposed, with NC9 providing the highest PR value (4.62 MPa) and NCo the lowest (3.50 MPa). Fontanela (2012) also found no significant differences in the PR below the 0.30 m layer in Argissolo Vermelho Amarelo with a sandy loam texture as a result of the treatments used. According to Landau et aL (2009), maize roots are concentrated in the first 0.30 m, so the PR values found as a function of compaction levels could restrict or even prevent the growth of the crop's root system. Veen and Boone (1990) *apud* Albuquerque and Reinert (2001) reported that maize roots stopped growing when the PR reached 4.7 MPa. Tavares Filho et aL (2001) in a study carried out on a loamy purple latosol found average PR values in the 0.15 - 0.35 m layer of 3.91 and 4.09 MPa for the no-till and conventional systems respectively, values similar to those observed in this study, and much higher than those described in the literature as impeding growth. The authors concluded that these values did not restrict the root development of the maize, but did influence its morphology.

4.2 Agronomic characteristics of the crop

4.2.1 Plant height (PH), first ear insertion height (EHE) and stem diameter (SD).

There was no significant interaction ($p < 0.05$) between tillage implements and compaction levels for AP, APE and DC, but these factors were influenced by tillage implements and compaction levels separately. Table 7 shows the average results for AP, APE and DC as a function of tillage implement.

Table 7. F-test and means for Plant Height (PH), First Spike Insertion Height (SPI) and Stalk Diameter (SD) as a function of tillage implements and compaction levels.

Causes of Variation	Reviews		
	AP (m)	APE (m)	DC (mm)
IP1	1,81 a	0,99 a	19,3a
IP2	1,52 c	0.92 ab	20,7 a
IP$_3$	1,69 b	0,87 b	19,6a
Average	1,67	0,93	19,8
F-test			
IP	57,10*	12,59*	5,15*
NC	12,08*	7,32*	6,18*
IP x NC interaction	1,25^{ns}	1,99^{ns}	1,16^{ns}
C.V. (%) - IP	4,66	7,40	6,60
C.V. (%) - NC	8,82	7,68	10,61

Averages followed by the same letters in the columns do not differ by Tukey's test at 5% probability. IPi = plough harrow; IP2 = mouldboard plough; IP3 = scarifier. ns = not significant; * = significant at 5% probability; C.V. = Coefficient of variation (%); IP = Soil preparation implements; NC = Compaction levels.

IP1 provided the highest PA and was statistically superior to the other tillage implements. IP3 had a statistically higher average than IP2. A similar behaviour was observed for APE, where IP1 also provided the highest value; however, it did not differ statistically from IP2, which did not differ from IP3. Tukey's test did not show any differences between the tillage implements for DC. It is likely that the physical change in the soil caused by IP1 favoured the development of these characteristics in the maize plant. Carvalho et al. (2012) assessed the performance of maize hybrids in north-eastern Brazil and, for the same hybrid used in this study, found an average AP value of 2.05 metres and APE of 1.09 metres. Trein (1988) in a study carried out on red Argissolo found greater maize plant heights when the soil was prepared with a disc plough and harrow.

Using the average PR values found, it was possible to verify that AP, APE and DC showed a decreasing linear behaviour from the PR of 1.53 MPa (Figure 6), resulting in reductions of 19, 14 and 15%, respectively for AP, APE and DC when the PR changed from 1.53 to 3.33 MPa. The results found in this study corroborate Silva (1998) and Rossetti and Centurion (2013), who found that the growth of the aerial part of maize plants responded negatively to an increase in PR, and Freddi (2007) in a similar study on a medium-textured red latosol. The latter found linearly decreasing values for the same parameters from a PR of 1.65 MPa. Freddi (2007) also found a 15, 18 and 10 per cent reduction in AP, APE and DC, respectively, when the PR was increased from 0.32 to 1.83 MPa. Freddi et al. (2009), in a similar study on medium-textured red latosol, found a linear reduction of 8 and 13 per cent in AP and DC when the PR was increased from 0.87 to 2.15 MPa, respectively.

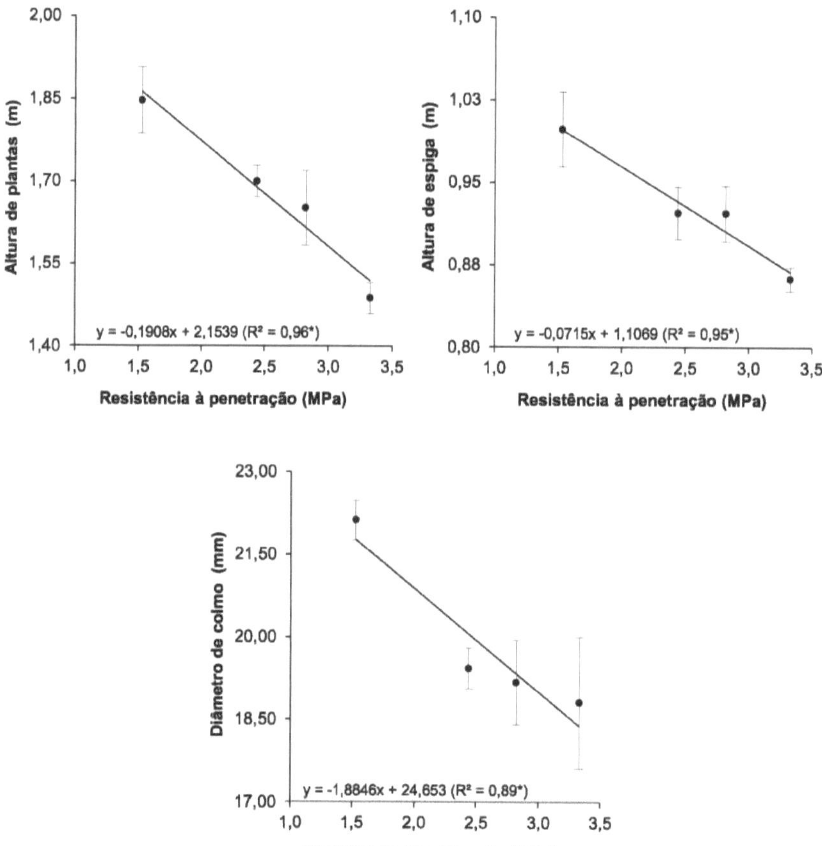

Figure 6 Regressions between the mechanical resistance of the soil to penetration of the Yellow Argissolo and plant height, height of insertion of the first ear and diameter of the maize stalk.

According to Letey (1985), an increase in PR causes reduced development of the root system which, by means of hormonal signals sent by the roots, reduces the development of the aerial part, thus potentially reducing productivity. According to Freddi (2007), taller plants with greater heights of insertion of the first ear are likely to achieve higher yields, as the reduction in vegetative mass causes a reduction in photosynthetic capacity, directly affecting grain production.

Although Sene et al. (1985) considered that, for sandy soils, only PR values above 6.0 MPa would be critical for crop development, the results found in this work show that for PA, APE and DC this value may be overestimated, as it was possible to see a reduction from a PR of 1.53 MPa, which is closer to the limit established by Taylor et al. (1966) and Silva et al. (1994) of 2.0 MPa.

4.2.2 Number of grains (NG), First ear rows (FE), First ear length (CE) and First ear diameter (DE).

There was no significant interaction ($p < 0.05$) between tillage implements and compaction levels for NG, FE, CE and DE, with only NG and FE being influenced by compaction levels (Table 8).

Table 8. F-test for Number of grains (NG), First ear row (FE), First ear length (CE) and First ear diameter (DE) as a function of tillage implements and compaction levels.

	Reviews			
Causes of Variation	NG	FE	EC	DE
F-test				
IP	$0,12^{ns}$	$4,75^{ns}$	$1,99^{ns}$	$2,40^{ns}$
NC	$3,81^{*}$	$6,22^{*}$	$0,95^{ns}$	$1,71^{ns}$
IP x NC interaction	$0,88^{ns}$	$0,99^{ns}$	$0,71^{ns}$	$0,16^{ns}$
C.V. (%) - IP	24,35	5,65	9,06	5,20
C.V. (%) - NC	19,14	6,51	9,82	5,40

ns = not significant; * significant at 5% probability; C.V. = Coefficient of variation (%). IP = Soil preparation implements; NC = Compaction levels.

The lack of significance for EC and ED may indicate that these are characteristics intrinsic to the genotype used and are generally influenced by fertiliser management (LOURENTE et aL, 2007; LOPES et aL, 2010) and plant population (VIEIRA et aL, 2010; BRACHTVOGEL, 2008). As the attributes AP, APE and DC were negatively affected by RP, it would be expected that CE and DE would also show similar behaviour, as discussed by Freddi (2007). In addition, NG and FE showed, respectively, quadratic and linear decreasing behaviour from a PR of 1.53 MPa, causing reductions in these factors of 20 and 11% respectively when the PR went from 1.53 to 3.33 MPa (Figure 7), again corroborating the results found by Freddi (2007).

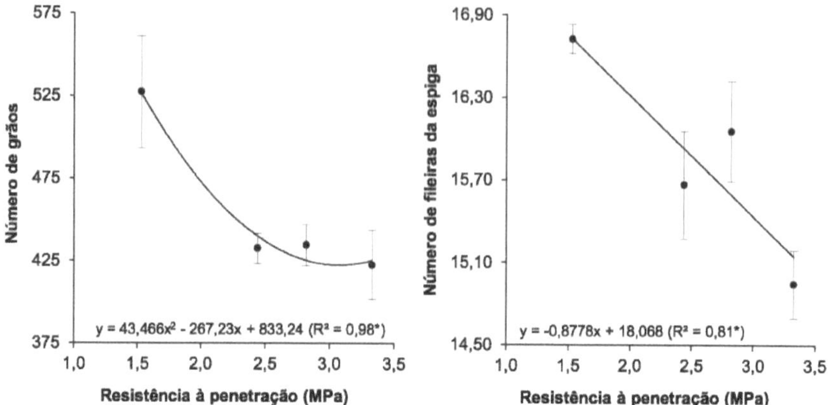

Figure 7. Regressions between the mechanical resistance of the soil to penetration of the Yellow Argissolo and the number of grains and rows of the first ear of maize.

As observed for the other crop growth parameters, the increase in PR caused by tractor traffic reduced NG and FE, which are important components of maize crop productivity. This fact can again be explained by Letey (1985), possibly because the increase in PR caused reduced development of the root system which, through hormonal signals sent by the roots, reduces the development of the aerial part, thus reducing productivity.

4.2.3 Plant dry matter (PDM) and root dry matter (RSM).

There was no significant interaction (p<0.05) between tillage implements and compaction levels for MSP and MSR. Both parameters were only influenced by compaction levels (Table 9).

Table 9. F-test for Plant Dry Matter (PDM) and Root Dry Matter (RSM) as a function of tillage implements and compaction levels.

Causes of Variation	Reviews	
	MSP	MSR
F-test		
IP	$0{,}84^{ns}$	$1{,}85^{ns}$
NC	$14{,}65^*$	$3{,}74^*$
IP x NC interaction	$1{,}83^{ns}$	$0{,}52^{ns}$
C.V. (%) - IP	20,18	29,57
C.V. (%) - NC	18,69	49,18

ns = not significant; * significant at 5% probability; C.V. = Coefficient of variation (%). IP = Soil preparation implements; NC = Compaction levels.

MSP showed a decreasing linear behaviour as a result of the increase in PR from 1.53 MPa, with a 39% reduction when PR went from 1.53 to 3.33 MPa (Figure 8). Similar results were found by Freddi (2007), who found a linear reduction of 24% in the MSP of the maize crop in a clay-textured red latosol when the PR went from 0.32 to 1.83 MPa. Freddi et al. (2009) also found a 26% linear reduction in the MSP of the maize crop in a medium-textured red latosol when the PR went from 0.87 to 2.15 MPa. Foloni et al. (2003) found a reduction of approximately 20% in the aerial growth of plants after 40 days of cultivation under 1.4 MPa of mechanical impedance in medium-textured red latosol. Again, the results presented here disagree with the critical limit established by Sene et al. (1985) of 6.0 MPa for sandy soils.

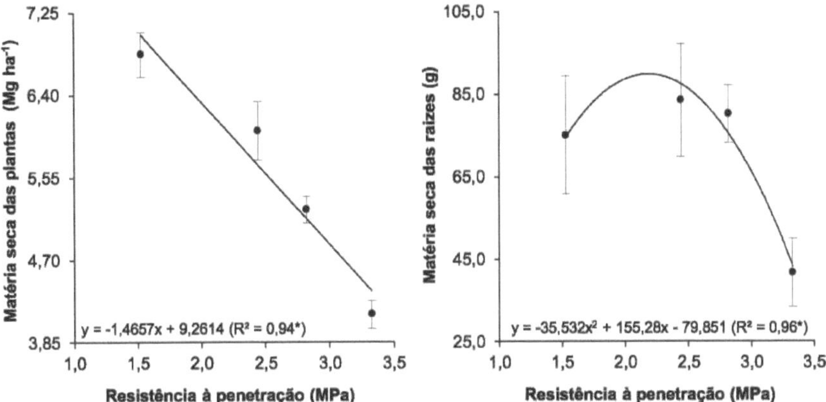

Figure 8. Regressions between the mechanical resistance of the soil to penetration of the Yellow Argissolo and the dry matter of maize plants and roots.

MSR showed a quadratic behaviour as a function of the increase in PR. It was possible to see that up to 2.18 MPa (maximum point of the curve obtained through the derivation of the quadratic equation) there was an increase in RSM and from this value onwards there was a reduction (Figure 8), a value much lower than the critical limit established by Sene et al. (1985) of 6.0 MPa for sandy soils and in agreement with that established by Taylor et al. (1966) and Silva et al. (1994) as critical to root and shoot development. Different results were found by Tavares Filho et al. (2001), where they found no restriction to the root development of the maize crop in a loamy purple latosol when the RP was 3.5 MPa.

The results shown in Figure 8 disagree with those obtained by Foloni et al. (2003), who found no change in MSR as a function of the increase in PR, leading them to conclude that MSR was not a suitable variable for gauging the sensitivity of maize plants to soil compaction. Freddi (2007), in a study on medium-textured red latosol, found the same quadratic behaviour; however, they found a reduction in MSR from a PR of 1.66 MPa to a PR of 3.09 MPa. For this author, from 3.09 MPa the root system responded to the restriction imposed by the soil with an increase in root diameter. In addition, this author observed an increase in MSR up to a PR of 5.69 MPa, which was not observed in the present study, where there was a reduction of approximately 53% when the PR went from

2.18 to 3.33 MPa. Freddi (2007) also found that MSR was a sensitive indicator of soil compaction.

Beutler and Centurion (2004a), working with rice crops in medium and clay loam, assessed MSR and observed behaviour similar to that found in this study. This slight compaction promotes closer contact between the soil, the solution and the roots, with an increase in the area of the soil explored by the roots so that the nutrients reach the absorption points more quickly.

4.2.4 Mass of 1000 grains (MG) and Productivity (P).

There was no significant interaction (p<0.05) between tillage implements and compaction levels for MG and P, with only P being influenced by compaction levels (Table 10).

Table 10. F-test for 1000-grain mass (MG) and productivity (P) as a function of soil preparation implements and compaction levels.

Causes of Variation	Reviews	
	MG	P
F-test		
IP	$1{,}17^{ns}$	$3{,}01^{ns}$
NC	$1{,}62^{ns}$	$5{,}09^{*}$
IP x NC interaction	$1{,}19^{ns}$	$0{,}44^{ns}$
C.V. (%) - IP	18,39	24,86
C.V. (%) - NC	15,32	17,77

ns = not significant; * significant at 5% probability; C.V. = Coefficient of variation (%). IP = Soil preparation implements; NC = Compaction levels.

The P results found in this work were higher than those found by Carvalho et al. (2012), for the same hybrid in a network of trials in the northeast region, and the maximum yield (14.78 Mg ha[-1]) was found in non-trafficked soil. It is possible that the non-limiting availability of water throughout the cycle favoured the development of the crop.

It was possible to see reductions of 15, 20 and 22 per cent respectively in maize crop yields when comparing non-trafficked soil to soil subjected to 3, 6 and 9 passes of a 3.5 Mg tractor (Figure 9). Tractor traffic increased PR, reducing root development and consequently the absorption of nutrients, especially phosphorus and potassium, which are preferentially transported in the soil by diffusion. Beutler et al. (2009) found a similar situation in a study carried out on arenic Yellow Red Argisol where machine traffic (8 passes of an 8 Mg tractor) reduced maize yields by 22% when compared to non-trafficked soil. Freddi (2007) found a 38% reduction in maize yields in medium-textured Red Latosol when the soil was subjected to 6 passes of an 11 Mg tractor when compared to non-trafficked soil.

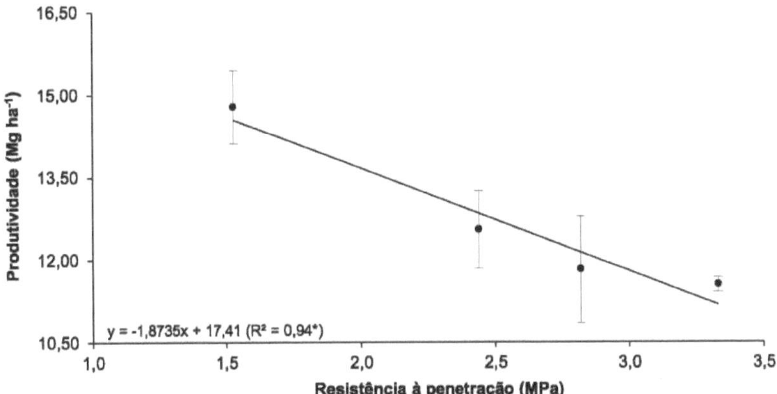

Figure 9. Regression between soil mechanical resistance to penetration of Yellow Argissolo and maize yield.

Figure 9 shows that the increase in PR values from 1.53 MPa onwards linearly reduced the productivity of the maize crop. This value is lower than that proposed by Taylor et al. (1966) and Silva et al. (1994) and much lower than that proposed by Sene et al. (1985) as limiting the development of crops in sandy soils. However, a value close to that found in this study was verified by Freddi (2007), where there was a quadratic reduction in the productivity of the maize crop in a medium-textured red latosol from a PR of 1.65 MPa. An even lower value was found by Beutler et al. (2009) in arenic Yellow Red Argisol, where there was a reduction in maize yields from a PR of 0.91 MPa. Mahl et al. (2008), on the other hand, did not find a reduction in maize yield at PR values close to 3.0 MPa. Freddi et al. (2009) found a significant reduction in maize yield in medium-textured red latosol when the PR was 2.15 MPa.

As can be seen, obtaining a critical PR value that limits crop development and productivity is quite difficult, since this physical attribute of the soil is extremely dependent on the water content, texture and structural condition of the soil. With an increase in humidity, PR can quickly change from a probable limiting condition to a non-limiting one (ROSSETTI and CENTURION, 2013). In addition to the factors intrinsic to the soil, crop responses are also different.

CHAPTER 5

CONCLUSION

Under the conditions in which this work was carried out, we conclude:

Soil preparation implements influenced density and total porosity in the 0.00 - 0.10 m layer. The plough harrow provided soil density values that were not critical to crop development, while total porosity showed higher values when the soil was prepared with the scarifier. The soil preparation implements did not affect the soil's mechanical resistance to penetration.

The soil preparation implements affected plant height, first ear insertion height and stalk diameter. Higher values for plant height and first ear insertion height were observed when the soil was prepared with the plough harrow.

The levels of compaction caused by tractor traffic led to an increase in soil density and total porosity in the 0.00 -0.10 m layer and an increase in the soil's mechanical resistance to penetration up to the 0.30 m depth.

Increasing the soil's mechanical resistance to penetration above 1.53 MPa restricted the following agronomic characteristics of the maize crop: plant height, height of insertion of the first ear, stalk diameter, number of grains, rows of the first ear, yield and plant dry matter. The dry matter of the roots increased up to the mechanical resistance of the soil to penetration of 2.18 MPa and then decreased.

CHAPTER 6

BIBLIOGRAPHICAL REFERENCES

ABREU, S. L.; REICHERT, J. M.; REINERT, D. J. Mechanical and biological scarification to reduce compaction in sandy loam under no-tillage. **Revista Brasileira de Ciência do Solo,** v.28, p.519-531, 2004.

ALBUQUERQUE, J. A.; SANGOI, L.; ENDER, M. Effects of crop-livestock integration on soil physical properties and maize crop characteristics. **Revista Brasileira de Ciência do Solo,** v.25, p. 717-723, 2001.

ALVARENGA, R. C.; COSTA, L. M.; MOURA FILHO, W.; REGAZZI, A. J. Legume root growth in artificially compacted soil layers. **Revista Brasileira de Ciência do Solo,** v.20, p.319-326, 1996.

ALVARENGA, R. C.; CRUZ, J. C.; NOVOTNY, E. H. **Maize cultivation: conventional soil preparation.** Sete Lagoas, MG: Embrapa Milho e Sorgo, 2002. 4p. (Embrapa Milho e Sorgo. Comunicado Técnico, 40).

AMARAL, F. C. S.; SILVA, E. F.; MELO, A. S. **Pedological characterisation and soil water infiltration studies in irrigated perimeters in the São Francisco Valley.** Rio de Janeiro, RJ: Embrapa Solos, 2006. 104p. (Embrapa Solos. Research and Development Bulletin, 97).

ANDRADE, C. L. T.; ALBUQUERQUE, P. E. P.; BRITO, R. A. L.; RESENDE, M. **Viabilidade e manejo da irrigação da cultura do milho.** Sete Lagoas, MG: Embrapa Milho e Sorgo, 2006. 12p. (Embrapa Milho e Sorgo. Circular Técnica, 85).

BALBINO, L. C.; OLIVEIRA, E. F. Effect of tillage systems on grain yields of wheat, soya and maize. In: BRAZILIAN CONGRESS OF AGRICULTURAL ENGINEERING, 20, 1991, Londrina. **Proceedings...** SBEA, 1992. p.1354-1360.

BENTIVENHA, S. R. P.; GONÇALVES, J. L. M.; SASAKI, C. M. Soil mobilisation and initial growth of eucalyptus as a function of the type of tine, working depth and soil characteristics. **Engenharia Agrícola,** Jaboticabal, v.23, n.3, p.588-605, 2003.

BERGAMIN, A. C.; VITORINO, A. C. T.; LEMPP, B.; SOUZA, C. M. A.; SOUZA, F. R. Root anatomy of maize in compacted soil. **Pesquisa Agropecuária Brasileira,** v.45, n.3, p.299-305, 2010
BEUTLER, A. N.; CENTURION, J. F.; MENGATTO, L. H.; ALVES, J. B.; WAGNER, G. P. C. Impact of machine traffic on soil physical quality and maize productivity in Argissolo. **Acta Scientiarum,** v.31, n.2, p.359-364, 2009.

BEUTLER, A. N.; CENTURION, J. F.; SILVA, A. P. Comparison of penetrometers in the evaluation of the compaction of latosols. **Engenharia Agrícola,** v.27, p.146-151, 2007.

BEUTLER, A.N.; CENTURION, J. F.; CENTURION, M. A. P. C.; LEONEL, C L.; SÃO JOÃO, A. C. G.; FREDDI, O. S. Optimum water interval for monitoring the compaction and physical quality of a red latosol cultivated with soya.
Brazilian Journal of Soil Science, v.31, p. 1223-1232, 2007.

BEUTLER, A.N.; CENTURION, J. F.; CENTURION, M. A. P. C.; SILVA, A. P. Effect of compaction on the productivity of soya cultivars in red latosol **Revista Brasileira de Ciência do Solo,** v.30, p.787-794, 2006

BEUTLER, A.N.; CENTURION, J.F. Resistência à penetração em latossolos: Valor limitante à produtividade de arroz de sequeiro. **Ciência Rural,** v.34, p.1793-1800, 2004a.

BEUTLER, A. N.; CENTURION, J. F. Dry matter and plant height of soya and rice as a function of compaction degree and water content of two latosols. **Engenharia Agrícola,** v.24, n.1, p. 142-149, 2004b.

BEUTLER, A. N.; CENTURION, J. F.; SILVA, A. P. Optimum water interval and soya and rice production in two Latosols. **Irriga,** v.9, p. 181-192, 2004.

BOHM, W. Methods of studying root systems. **New York: Springer-Varlag,** 194p. 1979.

BOLLER, W.; KLEIN, V. A.; DALLMEYER, A. U. Sowing maize in soil under reduced tillage. **Revista Brasileira de Ciência do Solo,** v.22, p.123-130, 1998.

BOTTA, G.; JORAJURIA, C. D.; DRAGHI, L. Influence of the axle load, tire size and configuration, on the compaction of a freshly tilled clayey soil. **Journal Terramechanics,** v.39, n.1, p.47-54, 2002. BOUKOUNGA, J. C. **Physical-mechanical properties of a clay loam under different management systems and soil preparations.** 2009. 89p. Dissertation (Master's in Soil Science) - Federal University of Rio Grande do Sul, Porto Alegre.

BRAIDA, J. A. **Organic matter and plant residues on the soil surface and their relationship with soil mechanical behaviour under no-till.** 2004. 106p. Thesis (Doctorate in Soil Science) - Federal University of Santa Maria, Santa Maria.

BRAZIL. Ministry of Agriculture. **Exploratory survey - reconnaissance of soils in the state of Pernambuco.** Recife: SUDENE, 1973. 354p.

BRACHTVOGEL, E. L. **Maize population densities and arrangements and agronomic components.** 2008. 96p. Dissertation (Master's in Agronomy) - Universidade Estadual Paulista,

Botucatu.

GAMARA, R. K.; KLEIN, V.A. Scarification in no-till farming as a soil and water conservation technique. **Revista Brasileira de Ciência do Solo,** v.29, n.5, p.789-96, 2005.

CAMARGO, O. A.; ALLEONI, L.R.F. **Soil compaction and plant development.** Piracicaba, SP: ESALQ, 1997. 132p.

CAMARGO, O. A.; MONIZ, A. C.; JORGE, J. A. & VALADARES, J. M. A. S. **Methods of chemical, mineralogical and physical analysis of soils from the Campinas Agronomic Institute.** Campinas, SP: Instituto Agronómico de Campinas, 1986. 94p. (Agronomic Institute of Campinas. Technical Bulletin, 106).

CANARACHE, A. Penetr-a generalised semi-empirical model estimating soil resistance to penetration. **Soil and Tillage Research,** v.16, p.51-70, 1990.

CARVALHO, H. W. L.; PACHECO, C. A. P.; CARDOSO, M. J.; ROCHA, L. M. P.; OLIVEIRA, I. R.; BARROS, I.; TABOSA, J. N.; LIRA, M. A.; OLIVEIRA, E. A.; MACEDO, J. J. G.; NASCIMENTO, M. M. A.; SIMPLÍCIO, J. B.; COUTINHO, G. V.; BRITO, A. R. M. B.; TAVARES, J. A.; TAVARES FILHO, J. J.; RODRIGUES, C. S.; CASTRO, C. R.; MENESES, M. C.; OLIVEIRA, T. R. A.; GOMES, M. C. M.; MENEZES, V. M. M.; SANTANA, A. F. **Performance of maize cultivars in northeastern Brazil: 2010/2011 harvest.** Aracaju, SE: Embrapa Tabuleiros Costeiros, 2012. 33p. (Embrapa Tabuleiros Costeiros. Technical Communication, 122).

CARVALHO FILHO, A.; CENTURION, J. F.; SILVA, R. P.; FURLANI, C. E. A.; CARVALHO, L. C. C. Soil preparation methods: changes in soil roughness. **Engenharia Agrícola,** Jaboticabal, v.27, n.1, p.229-237, 2007.

CENTURION, J. F.; DEMATTÊ, J. L. I. Efeito de sistemas de preparo de solo nas propriedades físicas de um solo sob cerrado cultivado com soja. **Revista Brasileira de Ciência do Solo,** v.9, p.263-266, 1985.

CENTURION, J. F.; DEMATTÊ, J. L. I. Cerrado soil preparation system: effects on physical properties and maize crop. **Pesquisa Agropecuária Brasiliera,** v.23, n.2, p.315-324, 1992.

CFSEMG. **Recommendation for the use of correctives and fertilisers in Minas Gerais 5° Approximation.** Viçosa, 1999. 359p.

CHAN, K. Y.; OATES, A.; SWAN, A. D.; HAYES, H. C.; DEAR, B. S.; PEOPLES, M. B. Agronomic consequences of tractor wheel compaction on a clay soil. **Soil Tillage Research,** v.89, p. 13-21, 2006.

CHOUDHURY, E. N.; MELLO, C. A. de O.; MORGADO, L. B.; **Soil preparation and residual fertilisation in maize cultivation in irrigated areas.** Petrolina, PE: Embrapa Semiárido, 1991.21 p. (Embrapa Semiárido. Boletim de Pesquisa, 40).

CINTRA, F. L. D.; MIELNICZUCK, J. Potential of some plant species for the recovery of soils with degraded physical properties. **Revista Brasileira de Ciência do Solo,** v.7, p. 197-201, 1983.

COLLARES, G. L. **Compaction in Latosols and Argissols and its relationship with soil and plant parameters.** 2005. 106p. Thesis (Doctorate in Soil Science) - Federal University of Santa Maria, Santa Maria.

COLLARES, G. L.; REINERT, D. J.; REICHERT, J. M.; KAISER, D. R. Soil physical quality on bean productivity in an Argissolo. **Pesquisa Agropecuária Brasileira,** v.41, p. 1663-1674, 2006.

CONAB. **Brazilian crop monitoring: grains, fourth survey, January 2014.** National Supply Company. Brasília: Conab, 2014. Available   at: <http://www.conab.gov.br/OlalaCMS/uploads/arquivos/14 01 10 15 07 19 bulletin _graosjaneiro_2014.pdf>. Accessed on: 08 Feb. 2014.
CORTEZ, J. W.; ALVES, A. D. S.; MOURA, M. R. D.; OLSZEVSKI, N.; NAGAHAMA, H. J. Physical attributes of yellow clay loam from the northeastern semi-arid region under tillage systems. **Revista Brasileira de Ciência do Solo,** v.35, n.4, p. 1207-1216, 2011.

CORSINI, P. C.; FERRAUDO A. S. Effects of cropping systems on soil density and macroporosity and on maize root development in purple latosol. **Pesquisa Agropecuária Brasiliera,** v.34, n.2, p.289-298, 1999.

COSTA, F. S.; ALBUQUERQUER, J. A.; BAYER, C.; FONTOURA, S. M. V.; WOBETO, C. Physical properties of a bruno loam affected by no-till and conventional tillage systems. **Revista Brasileira de Ciência do Solo,** v.27, p.527-535, 2003.

DE MARIA I. C.; CASTRO, O. M.; SOUZA DIAS, H. Soil physical attributes and root growth of soya beans in purple latosol under different tillage methods. **Revista Brasileira de Ciência do Solo,** v.23, p.703-709, 1999.

DIAS JÚNIOR M. S.; MIRANDA, E. E. V. Behaviour of the compaction curve of five soils in the Lavras-MG region. **Ciência Agrotécnica,** v.24, n.2, p.337-346, 2000.

DIAS JÚNIOR, M. S.; PIERCE, F. J. The soil compaction process and its modelling. **Revista Brasileira de Ciência do Solo,** v.20, p. 175-182, 1996.

DIAS JÚNIOR, M. S.; SILVA, A. R.; FONSECA, S.; LEITE, F. P. Alternative method for evaluating preconsolidation pressure using a penetrometer. **Revista Brasileira de Ciência do Solo,** v.28, p.805-810, 2004.

DORNELES, E. P. **Chemical attributes of clay loam and nutrient export by crops under tillage and fertiliser systems.** 2011. 91 p. Dissertation (Master's in Soil Science) - Federal University of Rio Grande do Sul, Porto Alegre.

BRAZILIAN AGRICULTURAL RESEARCH COMPANY. **Manual of soil analysis methods.** Rio de Janeiro, 2011.212p.

BRAZILIAN AGRICULTURAL RESEARCH COMPANY. **Brazilian soil classification system.** Brasilia, 2006. 370p.

FALLEIRO, R. M.; SOUZA, C. M.; SILVA, C. S. W.; SEDIYAMA, C. S.; SILVA, A. A; FAGUNDES, J. L. Influence of tillage systems on soil chemical and physical properties. **Brazilian Journal of Soil Science,** v.27, p. 1097-1104, 2003

FASINMIRIN, J. T.; REICHERT, J. M. Conservation tillage for cassava *(Manihot esculenta crantz)* production in the tropics. **Soil and Tillage Research,** v.113, n.1, p.1-10, 2011.

FEITOSA, J. R.; OLSZEVKI, N.; CORTEZ, J. W.; NAGAHAMA, H. J. Physical variables of yellow clay loam in the northeastern semiarid region as a function of periodic tillage operations. **Engenharia na Agricultura,** v.21, n.5, p.456-464, 2013.

FERREIRA, D. F. Statistical analyses using SISVAR (System for Analysis of Variance) for Windows 4.0 In: ANNUAL MEETING OF THE BRAZILIAN REGION OF THE INTERNATIONAL SOCIETY OF BIOMETRY, 45, 2000, São Carlos. **Proceedings...** UFSCar, 2000. p.255-258.

FOLONI, J. S. S.; CALONEGO, J. C.; LIMA, S. L. Effect of soil compaction on aerial and root development of maize cultivars. **Pesquisa Agropecuária Brasileira,** v.38, n.8, p.947-953, 2003.

FONTANELA, E. **Preparations and physical properties of a sandy soil for sugar cane and cassava in rio grande do sul.** 2012. 158p. Thesis (Doctorate in Soil Science) - Federal University of Santa Maria, Santa Maria.

FREDDI, O, S. **Evaluation of the optimum water interval in Red Latosol cultivated with maize.** 2007. 105p. Thesis (Doctorate in Agronomy) - Universidade Estadual Paulista, JaboticabaL

FREDDI, O. S.; CENTURION, J. F.; DUARTE, A. P.; LEONEL, C. L. Soil compaction and production of maize cultivars in red latosol: I - plant and soil characteristics and S index. **Revista Brasileira de Ciência do Solo,** v.33, n.4, p. 793- 803, 2009.

GAGGERO, M. R. **Changes in the physical and mechanical properties of soil under tillage and grazing systems.** 1998. 124p. Dissertation (Master's in Soil Science) - Federal University of Rio Grande do Sul, Porto Alegre.

GAMEIRO, C. A.; GABRIEL FILHO, A. Incoporation of plant remains, development and grain yield of maize (Zea *mays* L.) evaluated in two types of soils prepared by five plough models. In: CONGRESS
BRASILEIRO DE ENGENHARIA AGRÍCOLA, 28, 1999, Santa Maria. **Proceedings...** Santa Maria: RS: SBEA, 1999, CD-ROM.

GALETI, P.A. **Agricultural Mechanisation.** Campinas: Instituto Campineiro de Ensino Agrícola. 1988. 220p.

GILL, W. R.; VANDEN BERG, G. E. **Soil dynamics in tillage and traction.** Agricultural Research Service, United States Department of Agriculture, Washington, 1968, 511p.

GRIFFITH, D. R.; MANNERING, J. V.; GALLOWAY, H. M.; PARSONS, S. D.; RICHEY, C. B. Effect of eight tillage-planting systems on soil temperature, percent stand, plant growth and yield of corn on five Indiana Soils. **Agronomy Journal.** Madison, v.65, p.321-326, 1973.

GUIMARÃES, C. M.; STONE, L. F.; MOREIRA, A. A. J. Soil compaction in bean cultivation. II: effect on root and shoot development.
Brazilian Journal of Agricultural and Environmental Engineering, v.6, p.213-218, 2002.

YOKOYAMA, L. P.; SILVEIRA, P. M.; STONE, L. F. Profitability of maize, soya and wheat crops in different tillage systems. **Pesquisa Agropecuária Tropical,** v.32, n.2, p.75-79, 2002.

IMHOFF, S.; SILVA, A. P.; DIAS JÚNIOR, M. S.; TORMENA, C. A. Quantification of critical pressures for plant growth. **Revista Brasileira de Ciência do** Solo, v.25, p.11-18, 2001.

IMHOFF, S. D. C. **Indicators of structural quality and trafficability of red latosols and argissols.** 2002. 94p. Thesis (Doctorate in Agronomy) - Escola Superior de Agricultura "Luiz de Queiroz" Universidade de São Paulo, Piracicaba.

INOUE, G. H. Soil preparation systems and no-till farming in Brazil. **Agropecuária técnica,** v.24, n.1, p.1-11, 2003.

KAISER, D. R. **Structure and water in Argissolo under different preparations for maize cultivation.** 2010. 150p. Thesis (Doctorate in Soil Science) - Federal University of Santa Maria, Santa Maria.

KAMI MU RA, K. M. **Soil, machine and plant parameters as a function of plant residue doses and fertiliser deposition depths in direct sowing.** 2008. 129p. Dissertation (Master's in Soil Science) - Federal University of Rio Grande do Sul, Porto Alegre.

KLEIN, V. A. **Physical-hydraulic-mechanical properties of a purple latosol under different use** and management systems. 1998. 150p. Thesis (Doctorate in Soils and Plant Nutrition) - Escola Superior de Agricultura "Luiz de Queiroz", Universidade de São Paulo, Piracicaba.

KLEPKER, D. **Nutrients and roots in the profile and growth of maize and oats as a function of soil preparation and fertilisation methods.** 1991. 117p. Dissertation (Master's Degree in Soil Science) - Federal University of Rio Grande do Sul, Porto Alegre.

KLUTHCOUSKI, J.; FANCELLI, A. L.; DOURADO-NETO, D.; RIBEIRO, C. M.; FERRARO, L. A. Soil management and the yield of soya, maize, beans and rice in no-till farming. **Scientia Agrícola,** v.57, n.1, 2000.

LANÇAS, K. Subsoiling or scarification. **Cultivar Máquinas.** Available at: <http://www.qrupocultivar.com.br/site/content/artiqos/artiqos.php?id=416>. Accessed on: 10 October 2013.

LANDAU, E. C.; SANS, L. M. A.; SANTANA, D. P. **Maize cultivation: Climate and soil.** Embrapa Maize and Sorghum, 2009. Available at :< http://www.cnpms.embrapa.br/publicacoes/milho 5 ed/climaesolo.htm>. Accessed on: 22 December 2013.

LETEY, J. Relationship between soil physi cal properties and crop productions. **Advances in Soil Science,** v.1, p.277-294, 1985.

LIMA, C. L. R. **Soil compressibility versus traffic intensity in an orange orchard and animal trampling on irrigated pasture.** 2004. 70p. Thesis (Doctorate in Agronomy) - Escola Superior de Agricultura "Luiz de Queiroz", Universidade de São Paulo, Piracicaba.

LIMA, C. L. R.; SILVA, A. P.; IMHOFF, S.; LEÃO, T. P. Estimation of soil load-bearing capacity based on penetration resistance. **Revista Brasileira de Ciência do Solo,** v.30, p.217-223, 2006.

LOPES, M. M. S.; ALVES, G. A. R.; OLIVEIRA NETO, C. F.; OLIVEIRA, N. S.; JACKELINE, A. M.; SANTOS, D. G. C.; OKUMURA, R. S.; LOBATO, A. K. S.; WILSON, J. M.; MAIA, S. Cob length, diameter and dry matter in maize under the influence of various nitrogen levels. In: CONGRESSO NACIONAL DE MILHO E SORGO, 28, 2010, Goiânia. **CD-ROM.** Goiânia: Brazilian Maize and Sorghum Association, 2010. Available at :< http://www.abms.orq.br/cn milho/trabalhos/0188.pdf>. Accessed on: 09 Feb. 2014.

LOURENTE, E. R. P.; ONTOCELLI, R.; SOUZA, L. C. F.; GONÇALVEZ, M. C.; MARCHETTI, M. E.; RODRIGUES, E. T. Preceding crops, doses and sources of nitrogen on maize yield components. **Acta Scientiarum,** v.29, n.1, p.55-61, 2007.

MAHL, D.; SILVA, R. B.; GAMERO, P. R. A.; SILVA, P. R. A. Soil resistance to penetration, vegetation cover and maize productivity in no-till and scarified fields. **Acta Scientiarum,** v.30, n.5, p.741-747, 2008.

MAGALHÃES, P. C.; DURÃES, F. O. M. **Physiology of maize production.** Sete Lagoas, MG: Embrapa Milho e Sorgo, 2006, 10p. (Embrapa Milho e Sorgo. Circular Técnica, 76).

MAGALHÃES, P. C.; DURÃES, F. O. M.; PAIVA, E. **Physiology of the maize plant.** Sete Lagoas, MG: Embrapa Milho e Sorgo, 1994. 27p. (Embrapa Milho e Sorgo. Circular Técnica, 20).

MAZURANA, M.; LEVIEN, R.; MULLER, J.; CONTE, O. Tillage systems: changes in soil structure and crop yield. **Revista Brasileira de Ciência do Solo,** v.35, n.4, p.1197-1206, 2011.

MELLO IVO, W. M. P.; MIELNICZUK, J. Influence of soil structure on the distribution and morphology of the maize root system under three tillage methods. **Revista Brasileira de Ciência do Solo,** v.23, p. 135-148, 1999.

MEROTO JÚNIOR, A.; MUNDSTOCK, C. M. Wheat root growth as affected by soil strength. **Revista Brasileira de Ciência do Solo,** v.23, p.197- 202, 1999.

MOURA, P. M.; BEZERRA, S. A.; RODRIGUES, J. J. V.; BARRETO, A. C. Effect of compaction in two soils of different textural classes on radish cultivation. **Revista Caatinga,** v.21, n.5, p.107-112, 2008.
OLIVEIRA, E. F. de; BAIRRÃO, J. F. M.; CARRARO, I. M. Effect of tillage systems on some physical characteristics and grain yields of soya beans and maize. In: ORGANISATION OF COOPERATIVES OF THE STATE OF PARANÁ. **Research results for the 1987/88 summer harvest.** Cascavel: OCEPAR, 1989. p.233-237.

OLIVEIRA, G. C.; DIAS JÚNIOR, M. S.; CURI, N.; RESCK, D. V. S. Compressibility of a Clay Red Latosol according to soil water tension, use and management. **Revista Brasileira de Ciência do Solo,** v.27, p.773- 781, 2003.

RANEY, W. A.; ZINGG, A. W. **Principies of tillage.** In: USDA Yearbook of Agriculture, Washington, 1957, p.277-81.

PENEDO, E. D. **Chemical attributes of clay loam and nutrient export by crops under tillage and fertiliser systems.** 2011. 91 p. Dissertation (Master's in Soil Science) - Federal University of Rio Grande do Sul, Porto Alegre.

RADFORD, B. J.; BRIDGE, B. J.; DAVIS, R. J.; MacGARRY, D.; PILLAI, U. P.; RICKMAN, I. F.; WALSH, P. A.; YULE, D. F. Changes in properties of a Vertisol and responses of wheat after compaction with harvester traffic. **Soil and Tillage Research,** v.54, p.155-170, 2000.

REICHERT, J. M.; REINERT, D. J.; BRAIDA, J. A. Soil quality and sustainability of agricultural systems. **Ciência & Ambiente Magazine,** v.27, p.29-48, 2003.

REICHERT, J. M.; SUZUKI, L. E. A. S.; REINERT, D. J. Soil compaction in agricultural and forestry systems: Identification, effects, critical limits and mitigation In: CERRETA, C. A.; SILVA, L. S.; REICHERT, J. M. Tópicos em ciência do solo. **Brazilian Society of Soil Science,** v.5, p.49-134, 2007.

REINERT, D. J.; ALBUQUERQUE, J. A.; REICHERT, J. M.; AITA, C.; ANDRADA, M. M. C. Critical limits of soil density for root growth of cover crops in Red Argissolo. **Revista Brasileira de Ciência do Solo,** v.32, p.1805-1816. 2008.

RICHART, A.; TAVARES FILHO, J.; BRITO, O. R.; LLANILLO, R. F.; FERREIRA, R. Soil compaction: causes and effects. **Semina: Ciências Agrárias,** v.26, n.3, p.312-344, 2005.
ROSA, D. R. **Soil-machine-plant relationship in a cultivated clay loam and under native grassland.** 2009. 109p. Thesis (Doctorate in Agricultural Engineering) - Federal University of Santa Maria, Santa Maria.

ROSOLEM, C. A.; VALE, L. S. R.; GRASSE, H. F.; MORAES, M. H. Root system and nutrition of maize as a function of liming and soil compaction. **Revista Brasileira de Ciência do Solo,** v.18, n.3, p.491-497, 1994.

ROSSETTI, K. V.; CENTURION, J. F. Management systems and physical-water attributes of a Red Latosol cultivated with maize. **Brazilian Journal of Agricultural and Environmental Engineering,** v.17, n.5, p.472-479, 2013.

SÁ, J. C .M. Nutrient recycling from crop residues and fertilisation strategy for grain production in the no-till system. In: SEMINAR ON THE NO-TILL SYSTEM AT UFV. Viçosa, 1998. **Summary of lectures.** Viçosa, Federal University of Viçosa, 1998. p.19-61.

SALIRE, E. V.; HAMMEN, J. E.; HARDCASTLE, J. H. Compression of intact subsoils under short-duration loading. **Soil and Tillage Research,** v.31, p.235-248, 1994.

SCHAEFER, C. E. G. R.; SOUZA, C. M.; VALLEJOS, M. F. J.; VIANA, J. H. M.; GALVÃO, J. C. C.; RIBEIRO, L. M. Porosity characteristics of a Red-Yellow Argissolo submitted to different tillage systems. **Revista Brasileira de Ciência do Solo,** v.25, p.765-769, 2001.

SECCO, D. **Compaction states of two Latosols under no-till farming and their implications for mechanical behaviour and crop productivity.**
2003. 108p. Thesis (Doctorate in Soil Science) - Federal University of Santa Maria, Santa Maria.

SECCO, D,; REINERT, D. J.; REICHERT, J. M.; SILVA, V. R. Physical attributes and grain yield of wheat, soya and maize in two compacted and scarified Latosols. **Ciência Rural,** v.39, n.1, p.58-64, 2009

SEIXAS, J. **Levels of compaction in the cultivation of maize (Zea *mays*).** 2001. 80p. Dissertation (Master's Degree in Agronomy) - Federal University of Paraná, Curitiba.

SENE, M.; VEPRAAKAS, M. J.; NADERMAN, G. C.; DENTON, H. P. Relationships of soil texture and structure to corn yield response to subsoiling. **Soil Science Society of America Journal,** v.49, n.2, p.422-427, 1985.

SILVA, A. P. **Physical quality of the soil and the development of maize plants.** 1998. 80f. Tese (Livre Docência) - Escola Superior de Agricultura "Luiz de Queiroz", Universidade de São Paulo, Piracicaba.

SILVA, A. P.; KAY, B. D.; PERFECT, E. Characterisation of the least limiting water range of soils. **Soil Science Society of America Journal,** v.58, p. 1775-1781, 1994.

SILVA, A. P.; KAY, B. D. Estimating the least limiting water range of soils from properties and management. **Soil Science Society of America Journal,** v.61, p.877-883, 1997.

SILVA, A. P.; TORMENA, C. A.; IMHOFF, S. Optimum water interval. In: MORAES, M. H.; MULLER, M. M. L.; FOLONI, J. S. S. **Physical quality of the soil: methods of study-soil preparation and management systems.** Jaboticabal: Funep, p.1-18, 2002.

SILVA, A. P. Soil physics - LSO 0310. In: BRADY, N. C.; WELL, R. R. **The nature and Properties of soils.** 13. ed 2008

SILVA, S. R.; BARROS, N. F.; VILAS BOAS, J. E. B. Growth and nutrition of eucalyptus in response to compaction of latosols with different moistures. **Revista Brasileira de Ciência do Solo,** v.30, p.759-768, 2006.

SILVA, V. **Physical and water properties of soils under different states of compaction.** 2003. 171 p. Thesis (Doctorate in Agronomy) - Federal University of Santa Maria, Santa Maria.

SILVA, V.; REINERT, D. J.; REICHERT, J. M. Soil density, chemical attributes and root system of maize affected by grazing and soil management. **Revista Brasileira de Ciência do Solo,** v.24, n.1, p.191-199, 2000a.

SILVA, V.; REINERT, D. J.; REICHERT, J. M. Soil mechanical resistance to penetration influenced by harvester traffic in two soil management systems. **Ciência Rural,** v.30, n.5, p.795-801, 2000b.

SILVA, V. R.; REINERT, D. J.; REICHERT, J. M.; SOARES, J. M. Factors controlling the compressibility of an arenic dystrophic Red-Yellow Argisol and a typical dystrophic Red Latosol. I - Initial state of compaction. **Revista Brasileira de Ciência do Solo,** v.26, p.1-8, 2002.

SILVA, V. R.; REICHERT, J. M.; REINERT, D. J. Spatial variability of soil resistance to penetration in no-tillage. **Ciência Rural,** v.34, p.399-406, 2004.

STRECK, C. A. **Soil compaction and its effects on root development and productivity of bean and soya crops.** 2003. 83p. Dissertation (Master's Degree in Agronomy) - Federal University of Santa Maria, Santa Maria.

STRECK, C. A.; REINERT, D. J.; REICHERT, J. M.; KAISER, D. R. Changes in physical properties with soil compaction caused by induced tractor traffic in no-till farming. **Ciência Rural,** v.34, n.3, p.755-760, 2004.

STOLF, R.; FERNANDES, J.; FURLANI NETO, V. Impact Penetrometer - model IAA/Planalsucar - STOLF. **STAB,** v.1, p. 18-23, 1983.

STOLF, R. Theory and experimental test of formulas for transforming impact penetrometer data into soil resistance. **Revista Brasileira de Ciência do Solo,** v.15, p.229-235, 1991.

STOLF, R.; THURLER, A. M.; BACCHI, O. O. S.; REICHARDT, K. Method to estimate soil macroporosity and microporosity based on sand content and bulk density. **Brazilian Journal of Soil Science,** v.35, p.447-459, 2011.

STOLF, R.; SILVA, J.R.; GOMEZ, J.A.M. Agricultural harrows: 5- Historical evolution of their bearings. **ALCOOLbrás,** v.10, n.115, p.65-69, 2008.

STONE, L. F.; MOREIRA, J. A. A. Effects of tillage systems on soil compaction, water availability and bean behaviour. **Pesquisa Agropecuária Brasileira,** Brasília, v.34, n.1, p.83-91, 1999.

SUZUKI, L. E. A. S. **Soil compaction and its influence on physical properties.** 2005. 149p. Dissertation (Master's Degree - Soil Science), Federal University of Santa Maria, Santa Maria.

TAVARES FILHO, J.; BARBOSA, G. M. C.; GUIMARÃES, M. F.; FONSECA, I. C. B. Soil resistance to penetration and development of the root system of maize (Zea *mays)* under different management systems in a purple latosol. **Revista Brasileira de Ciência do Solo,** v.25, p.725-730, 2001.

TAYLOR, H. M., ROBERSON, G. M.; PARKER, J. J. Soil strength - root penetration relations to medium to coarse - textured soil materials. **Soil Science,** v.102, p. 18-22, 1966.

TORMENA, C. A.; SILVA, A. P.; LIBARDI, P. L. Characterisation of the optimum water interval of a purple latosol under no-tillage. **Revista Brasileira de Ciência do Solo,** v.22, p.573-581, 1998.

TORMENA, C. A.; SILVA, A. P.; LIBARDI, P. L. Soil physical quality of a Brazilian Oxisol under two tillage Systems using the least limiting water range approach. **Soil Tillage Research,** v.52, p.223-232, 1999.

TORRES, E.; ODILON, F. S.; GALERANI, P. R. **Soil management for soya cultivation.** Londrina, PR: Embrapa Soja, 1993. 71p. (Embrapa Soja. Technical Circular, 12).

TREIN, C. **Soil preparation methods for maize cultivation and clover reseeding, in the oats+clover/maize rotation, after intensive grazing.** 1988. 111 p. Dissertation (Master's Degree - Soil Science) - Federal University of Rio Grande do Sul, Porto Alegre.

VIANA, J. H. M.; CRUZ, J. C.; ALVARENGA R. C.; SANTANA, D. P. **Soil management for maize cultivation.** Sete Lagoas, MG: Embrapa Milho e Sorgo, 2006, 14p. (Embrapa Milho e Sorgo. Circular Técnica, 77).

VIEIRA, M. A.; CAMARGO, M. K.; DAROS, E.; ZAGONEL, J.; KOEHLE, H. S. Maize cultivars and plant population affecting green ear yield. **Acta Scientiarum,** v.32, n.1, p.81-86, 2010.

CHAPTER 7

APPENDIX

APPENDIX A - Photographs of the experiment, soil preparation, compaction levels, sample collection and crop phases.

Valtra 785 TDA tractor during soil preparation with plough harrow.

ATTORNEY plough harrow.

Experimental area after 0 harrow preparation.

Valtra 785 TDA tractor during soil preparation with a mouldboard plough.

AST subsoiler.

Valtra 785 TDA tractor during levelling.

Valtra 785 TDA tractor during compaction application.

Greenhouse.

Samples for determining Ds and Pt.

Maize crop at stage V1.

Maize crop at the V4-V5 stage.

Top dressing.

Maize crop at stage V10-
V12.

Differences in plant height caused by the

treatments.

Printed by Books on Demand GmbH, Norderstedt / Germany